PRAISE FOR

Flying Blind

"So what happened to the idealistic '60s youth who went back to the land? *Flying Blind* is one couple's answer. Don Mitchell presents a rich, evocative account of wise stewardship—and of how making ends meet on a Champlain Valley farm in Vermont becomes a conservation success story in the fight to save endangered Indiana bats."

—Andrew Walker, executive director,
Bat Conservation International

"Don Mitchell has written a classic story of Vermont, of family, of farming, and of the evolving, never romantic, always crucial story of the encounter between people and the larger world."

—Bill McKibben, author of *Oil and Honey: The Education of an Unlikely Activist*

"Don Mitchell's *Flying Blind* does for rural New England what Wendell Berry's essays do for Kentucky and Norman Maclean's *A River Runs through It* does for the American West. On one level, *Flying Blind* tells the engaging and often hilarious story of a man's determination to make his upcountry Vermont farm a welcoming home for an endangered and much-maligned species of 'flying rat.' It's also the story of how place, the past, family, and meaningful work can still form character at a time when much of America is increasingly alienated from nature, history, and community. Beautifully written, relentlessly honest, and unfailingly entertaining, *Flying Blind* is the book Don Mitchell was born to write."

—Howard Frank Mosher, author of *The Great Northern Express*, *Walking to Gatlinburg*, and *On Kingdom Mountain*

"Receiving a government grant to control invasive plants in the bat habitat around his farm was just the beginning. Don Mitchell hilariously chronicles the official visits and requirements that soon became such a prominent part of his life, along with the stupefying labor involved in grubbing up all that garlic mustard. What makes *Flying Blind* such a remarkably powerful memoir is Don Mitchell's quest to connect the ecological puzzle of bats' susceptibility to white-nose syndrome with his passionate and lifelong resistance to authority. At the deepest level, this is a story about how forgiveness and celebration help him find a trail through the woods to family and home."

—John Elder, author of *The Frog Run* and
coeditor of *The Norton Book of Nature Writing*

"In *Flying Blind*, Don Mitchell not only gives us a wonderful story about creating habitat for bats on his land, but tells about his own personal journey of becoming a bat-loving conservationist. In addition to the many scientific bat-conservation efforts taking place around the world, we also need stories like this—of an individual developing a greater understanding of bats, and of the natural world, and coming away better for it."

—Merlin Tuttle, founder,
Bat Conservation International

"*Flying Blind* is Don Mitchell's surprising story of how a vague fear of bats and a deep-seated mistrust of government inspired him to far greater intimacy with and stewardship of his one hundred and fifty acres in Vermont. It's a story about learning to love, learning to trust, and learning to make peace with your past. Read and enjoy."

—Sy Montgomery, author of *Journey of the Pink Dolphins*
and *Walking with the Great Apes*

FLYING BLIND

Also by Don Mitchell

Fiction
Thumb Tripping
Four-Stroke
The Souls of Lambs
The Nature Notebooks

Essays
Moving UpCountry
Living UpCountry
Growing UpCountry

Travel
Vermont: A Compass American Guide

FLYING BLIND

One Man's *Adventures* Battling Buckthorn, Making *Peace with* Authority, *and* Creating a *Home for* Endangered Bats

DON MITCHELL

Chelsea Green Publishing
White River Junction, Vermont

Jacket and title page illustrations:
Bat illustration by Ernst Haeckel, *Kunstformen der Natur (Artforms of Nature)* (1904); buckthorn illustration by William Dwight Whitney, *The Century Dictionary and Cyclopedia: An Encyclopedic Lexicon of the English Language* (New York, NY: The Century Co., 1889); shagback hickory illustration by Charles Sprague Sargent, *Manual of the Trees of North America* (Boston, MA: Houghton Mifflin Company, 1905).

Project Manager: Bill Bokermann
Developmental Editor: Brianne Goodspeed
Copy Editor: Ellen Brownstein
Proofreader: Helen Walden
Designer: Melissa Jacobson

Printed in the United States of America.
First printing August, 2013.
10 9 8 7 6 5 4 3 2 1 13 14 15 16

Our Commitment to Green Publishing
Chelsea Green sees publishing as a tool for cultural change and ecological stewardship. We strive to align our book manufacturing practices with our editorial mission and to reduce the impact of our business enterprise in the environment. We print our books and catalogs on chlorine-free recycled paper, using vegetable-based inks whenever possible. This book may cost slightly more because we use recycled paper, and we hope you'll agree that it's worth it. Chelsea Green is a member of the Green Press Initiative (www.greenpressinitiative.org), a nonprofit coalition of publishers, manufacturers, and authors working to protect the world's endangered forests and conserve natural resources. *Flying Blind* was printed on Natures Book, a 30-percent postconsumer recycled paper supplied by Thomson-Shore.

Library of Congress Cataloging-in-Publication Data
Mitchell, Don, 1947-
Flying blind : one man's adventures battling buckthorn, making peace with authority, and creating a home for endangered bats / Don Mitchell.
pages cm
ISBN 978-1-60358-520-0 (hardback) — ISBN 978-1-60358-521-7 (ebook)
1. Bats—Conservation—Vermont. 2. Bats—Conservation—Champlain Valley. 3. Bats—Ecology—Vermont. 4. Endangered species—Vermont. 5. Wildlife watching—Vermont. 6. Wildlife habitat improvement—Vermont. 7. Forest management—Vermont. 8. Vermont. Fish and Wildlife Department. 9. Mitchell, Don, 1947-—Homes and haunts—Vermont. 10. Mitchell, Don, 1947-—Philosophy. I. Title.
QL737.C5M678 2013
599.4—dc23

2013015613

Chelsea Green Publishing
85 North Main Street, Suite 120
White River Junction, VT 05001
(802) 295-6300
www.chelseagreen.com

In Memoriam

Wayne Treleven Mitchell

1915–2009

Contents

Author's Note

Bats are not blind, but the skill with which they fly owes not so much to vision as to echolocation. Akin to sonar navigation at sea, echolocation works by bats emitting a loud ultrasonic cry and then evaluating the sound that bounces back. Equipped with this means of perception and enormous wings (a bat's wing is actually an overgrown hand, with a membrane stretched between the bones of each finger), bats can turn on a dime in flight. They *need* this maneuverability in order to catch insects, which are also on the wing. This makes it hard for us, watching them wheel and turn, to fathom where they're apt to go from one moment to the next. The flight patterns of bats are seldom smooth and birdlike; often they are completely unpredictable. Surely this accounts for some measure of the fascination—and the fear, too—with which we tend to regard them.

The facts in this account are "true," so far as I understand the meaning of that word. But they are presented in a way that I hope will mimic a bat's peripatetic flight: bouncing from here to there, from one topic to the next without advance warning. Much of life is lived that way, looked at from a certain angle. Arguably, *most* of mental life is lived that way. And it seems appropriate to use such a strategy to narrate such a strange and unpredictable tale. If my readers ask "How did I get here?" from time to time, I hope they'll sense the writer's urge to veer off course suddenly (or dance off course, hopefully), to catch some bug that showed up on the radar unexpectedly. Writing, too, often takes place in something like the dark. Perhaps it can benefit from echolocation, too.

Habitat

Early in the summer of 2006, a biologist from Vermont's Fish and Wildlife Department asked if he could try his luck at trapping bats on our farm, which spreads across a picturesque corner of the Champlain Valley. Local boosters used to call our corner of the state—Addison County, with its Shire Town in Middlebury—"The Land of Milk and Honey," referencing a Biblical description of the Promised Land. That was before the price of milk went sour and the bees began experiencing Colony Collapse Disorder. Why Scott Darling chose our land to look for bats was not exactly clear to me, but the terrain has many features that attract the eye; this explains entirely why my wife, Cheryl, and I agreed to buy the place on the day we first saw it, forty years ago.

Our farm has the contours of a semiprivate valley, with rock walls that demarcate its boundaries to the south and east. A beaver swamp spreads across the base of the eastern cliff, from which a lazy stream drains the land into the Otter Creek. Much of the acreage consists of rolling fields of grass, used by us for pasturing a flock of sheep and making hay. But most of the land is in permanent forest, with handsome sugar maples predominant in stands that show a nice array of Northern hardwoods: red and white oak, ash, beech, hophornbeam, and hickory. Off to the west, on a clear day we can see the thin blue line of Lake Champlain—some fifteen miles distant—followed by a staircase of Adirondack peaks that rises up, row on row, toward Whiteface and Mount Marcy.

In a way, our farm might even serve as a proxy for the wider Champlain Valley in a naturalist's eye. If there were bats abroad in the landscape, why wouldn't they show up here? According to Darling, nine bat species are native to Vermont, and in 2006 only one of them—the Indiana bat, better known to residents of Indiana—had been listed as endangered by the U.S. Fish and Wildlife Service. Finding an Indiana bat on our premises would, I had to think, be an exciting coup.

The plan was to set up several "mist nets" in the evening at places where the bats might be apt to fly into them. Bats emerge from daytime roosts just as dusk is settling in, then fly around hunting insects all night long. Once tangled up in a mist net and extracted, a captured bat—with a fighting weight of, say, ten grams—could be fitted out with a miniscule radio transmitter and antenna for telemetric tracking. Darling would clip a patch of hair off the back of a detainee, then affix the transmitting apparatus with surgical glue. Weighing a mere four-tenths of a gram, the radio contraption came in just under the "five percent rule" for burdening a subject of investigation; once in place, it would transmit the bat's location till either the battery went dead or the glue failed. The data thus acquired would help to identify the trees where bats were roosting. Since female bats tend to gang up in maternity colonies to raise their pups, tracking one bat to its roost tree could lead to the discovery of many others. Sometimes several hundred of them, in a sort of bat hotel. That, in turn, could help researchers better understand the bats' habitat requirements.

I had no particular interest in bats back then; in fact, I would say I had a fairly strong aversion to them. On the few occasions when I'd actually seen a bat skitter through the night sky—flying with the crazy, unpredictable movements that call to mind the way a fox can dance across the land—my response was apprehensive. Flying rats, they seemed to me. They gave me a case of the willies. They could creep me out. For one thing, bats shunned the day and only came out at night. In Western culture, they had long been linked to witchcraft and occult activities. Satan had commonly been depicted having bat wings dating back at least as far as Milton and Dante. Then there was the widespread belief that bats were rabid. That they'd land on humans and get tangled in their hair. All things considered, bats did not have a good rep. But I like to think of myself as a friend of science. And I think I felt a twinge of pride—even cockiness—that the needs of wildlife biology, in the form of this investigation into Chiroptera, had taken an interest in the land we call our own. Sure, I said. Bring on the nets.

How Cheryl and I came to be in a position to let government agents trap bats on our farm is a story that could only have happened in the U.S.A.—and probably only in the era of our lifetimes. While attending college in the smoke-filled sixties, my girlfriend and I became like lemmings caught up in the cultural tides that were sweeping the land. When, in 1967, someone blew a whistle and directed all the high-minded, disaffected young people to head to San Francisco for a Summer of Love, we stuck our thumbs out and hitchhiked for a week or so to get in on that action. The following summer, when word came down that Boston was going to be the happening place—in order to foment a new American Revolution—we decamped from Swarthmore to a low-rent apartment on the back of Beacon Hill; there I wrote up a collection of stories based on hitchhiking adventures from the summer before. Life-as-lived, recast as the thinnest sort of fiction. Improbably, the book found a well-respected publisher within a few weeks of my graduating college—with a degree in philosophy, of all things. Clearly, being practical was not my strong suit. Then Peter Fonda's movie *Easy Rider* hit the screens, and the canny agent representing my material recognized that my stories suddenly had film potential. Four months later, my travelling companion and I had gotten married, bought a swank sports car, and moved to Los Angeles so I could work at MGM, writing a screenplay about reckless kids "thumb tripping" to see where the road would take them. Two years earlier, Cheryl and I had been standing at the freeway entrance; suddenly our ride was a canary-yellow Porsche.

Hollywood turned out not to be my cup of tea. For one thing, I don't like being told what to do—and a twenty-two-year-old screenwriter was destined to be told what to do by a wide array of colleagues and collaborators. Most of them were many years older than I was, and they were mostly men. Surrogate fathers, that is to say. I had no need for a surrogate father, since I had a real one judging my every move. As the film project progressed, I came to feel like the scribe to a committee rather than a glorified literary artist. Cheryl and I recognized, too, that there were contradictions between our professed countercultural values and the über-materialism of the film world. How to resolve this? I wrote my script, deposited some fat checks, and then started casting about for something else to do. Something more consistent with the person who I thought I was, or hoped I might become.

The *Whole Earth Catalog*, in one of its last large-format print editions, made the bold suggestion that we all move to Vermont—by all, I mean the army of rebellious kids whose pipe dreams now included growing vegetables, milking the odd goat, and building vernacular houses with their own hands. "Greening" America was our official goal. And to do it, we were going Back to the Land. Vermont, opined the author of this sidebar in the *Whole Earth Catalog*, was a very small state, so we could take it over quickly. Lots of farms were on the market, thanks to a shakeout in the dairy business. (This shakeout, I should note, is a perpetual phenomenon; over the last fifty years the number of Vermont dairies has dropped from over ten thousand farms to a scant nine hundred and change.) But the kicker was that all the dour, taciturn Vermonters would be likely to accept us, funny cigarettes and all. True, most of the natives were rock-ribbed Republicans—almost to a person, the writer assured—but their underlying values were libertarian, holding for each citizen's right to do his thing. So they ought to welcome long-haired hippies to their state.

I told Cheryl we should put some thought into moving to Vermont. We had both paid visits there during our college years; we had seen the white-steepled villages, the sugar houses puffing steam into the air. We longed for a four-season climate, after California. There had been some farming in my family on my father's side—mostly dairy farms, located mainly in Wisconsin—so the thought of trying my hand at agriculture did not seem preposterous. We began to seek out real estate listings and learn what land was selling for in different regions of the state. Not that we could move there yet, since I found myself indentured to the U.S. government for a couple dismal years. Within weeks of the day I would have been conscripted into the army—and, in all likelihood, sent to Vietnam—I convinced my draft board that they were dealing with a pacifist deserving C.O. status: an objector to war on grounds of conscience. Rather than getting a trip to southeast Asia, then, I signed on for two years of work deemed to be in the national interest—very broadly construed, in my case—in southeast Pennsylvania. And just as that stint was coming to an end, Cheryl and I spent a couple days in Vermont and saw the piece of property we came to call our own.

The afternoon was raw and rainy; fog hung like grey crepe against the cliffs that frame the farm on two of its borders. We walked partway into a grassy bowl—a broad, rolling meadow—and then turned around. We knew. Even on a rotten day, the place was irresistible. It was a rash and a reckless decision; that was who we were, though, in our early years of life together. So, with an impulsive nod and nothing like due diligence, we became the owners of a farm that was supposed to be one hundred fifty acres, more or less. "More or less" is a term of art in Yankee real estate transactions that gives the seller a ten percent leeway—plus or minus—in the parcel's actual size, which in principle could be determined by a survey if the buyer wanted to pay for it. We loved the land and feared that someone else would grab it if we waited on a survey, which could have taken months. So we left a small deposit—earnest money—with the mildly astonished seller and his wife, and a few weeks later we returned to close the deal. The year was 1972, and we were all of twenty-four years old.

Had someone from the government come to me in 1972 and asked permission to trap bats on our land—or do anything else, for that matter—I would have sent him packing. Most of my life had been a struggle with authority; halfway through my first day of school, for example, I got bored and left the building, making my way home by walking for a mile or so. It was during naptime, after milk and cookies. The teacher grew frantic when she noticed that my mat was empty; she went to the principal, a man who was not amused. Soon enough, the phone rang and my mother got an earful. How was I to know, though, that attending school was not a choice?

Gradually I came to accept the educational establishment's authority, though not without a measure of simmering resentment. Then there was the burden of authority represented by my family's Baptist church, whose strictures on behavior were supposedly enforced by God. The God of Moses—Yahweh. Or YHWH, for those who felt His name was too holy even to pronounce. Not the sort of god with whom a child wants to mess. There were Boy Scout leaders and Little League coaches; there were marching-band directors and camp counselors. I like to imagine that I chafed against all of them, struggling toward a version of adulthood that implied rebellion.

Then there was my father, who could also be a stern enforcer—probably the sternest of them all, in my encounters. He had a keen eye for good and bad child behavior, and he used both stick and carrot to make me *straighten up,* as he would say, *and fly right*. Another expression of his—this one a warning if I wasn't flying right—was *I'm going to take you on my knee and paddle you*. Mostly it was just a threat, but sometimes he made good on it. And for all I know I might have given him just cause. The odd thing was that my father never used a paddle; what I remember is the flat of his broad hand, calibrating a degree of force to suit my crime. Even to my childish mind, it gave rise to a thoughtful question: why calling it "paddling" if there wasn't any paddle? And perhaps I sensed that maybe *he* had his fanny paddled, somewhere along the line. Maybe by his own father.

But when it came to mistrusted authorities, government in all its forms was up there on my list. Way up. The nation's long war in Southeast Asia was a major factor, and it took me years to move beyond my sense of outrage. Not just for what we Americans did there—did to the Vietnamese—but over the drafting of young men like myself to fight and even give their lives there, whether they actually wanted to or not. And but for having attended a Quaker college, I would have been one of them. I had no interest in trying to make other people do things that they didn't want to; why were other people bent on bringing me to heel?

Almost as soon as I "went back to the land," though, government authority became cast in a different light. Every county had an agricultural office where bureaucrats were paid to see that folks like me succeeded at their efforts to make good at farming—and could throw some taxpayers' money in a farmer's direction if he would perform certain specified practices. Were your fields getting too acidic on the pH scale? Government would share the cost of having huge trucks come and drive across those needy acres, broadcasting lime. Were your fields sopping wet—too wet to put a tractor in—at a time of year when all your neighbors were out haying? Government had a bag of tricks to help you dry your fields. Were your livestock churning up mud and muck along the banks of an eroded stream? Government had money for riparian fencing.

On and on the list of helpful practices went, thanks to a perception that a poorly managed farm is going cost the nation plenty—either sooner or later—in topsoil losses and egregious pollution. But what really turned me on to working with the government was an incentive program tailored to reward those farmers who produced the nation's crop of wool. Woolgrowers, they called themselves with a quiet pride. And as new Vermonters, Cheryl and I had joined their ranks. In 1976 we acquired a flock of sheep and set out to care for them. Why? Because the *Whole Earth Catalog* had said we should. Sheep were easy keepers, we had read. Low capital requirements, low labor inputs. And they produced a yearly crop of not just lambs for meat, but also a specialty fiber known and loved by all. A guaranteed two-fer.

Back in those early years of amateur shepherding, you had to pay a specialist—a shearer, and there weren't that many—one or two bucks per head to come to your farm, set up a shearing rig, and give each sheep an annual fleecing. This is hair that grew on the animals' backs for free, just the way it does on the top of your head. An eight-pound fleece could be sold for up to twelve dollars in those halcyon days. Then the government would come along and pay you twenty more as a thank-you for having produced that fleece. The more you sold your wool for, the more the government would thank you—that's why it was called an "incentive" program. Why on earth? one might well ask. Well, the woolgrowers' lobby had convinced the federal government that wool was a strategic commodity, of which the nation needed to have an assured supply. Which meant, in effect, a *domestic* supply. The reason was that military uniforms demanded wool; no self-respecting army would send soldiers off to war clad in cheap synthetics. (Let us overlook, for now, the ironies involved in having pacifist shepherds growing wool for the armed forces. Let us also overlook how mohair, from angora goats, also found its way into the government's wool program; military uniforms do not require mohair.)

The rationale for uniforms with solid woolen content struck me as improbable, but I kept my mouth shut. The combination of a decent market price for wool plus the federal support program made it possible to feed a ewe through winter at a nominal profit, whether or not she delivered and raised a lamb. Now and then a murmur of dissent about the Wool Program would be voiced by someone in the halls of Congress, but the wool producers had an active set of advocates in the delegations from the intermountain states—where most of the nation's sheep were being raised. In fact, they were being raised on land that ranchers rented on the cheap from the government's Bureau of Land Management, a division of the Department of the Interior. Still, there were concerns that our wool checks would be targeted at some point for extinction. Then in 1982, *mirabile dictu*, Margaret Thatcher sent the British fleet to take back the Falkland Islands after an act of Argentinian aggression. When an Exocet missile struck and sunk the HMS *Sheffield*—a destroyer-class warship—in the South Atlantic, most of the casualties borne by British sailors were caused not by drowning or explosions *per se*, but by burns to men whose uniforms suddenly went up in flames and melted on their skin. Polyester uniforms, because the British navy was too cheap to spring for wool. America's hardworking shepherds were jubilant; our program of price supports now looked unassailable.

Actually, things got even better than that. A new category of largesse, the Unshorn Lamb Payment, found its way into the ledgers of our enterprise. When lambs are sold for slaughter, they have a modest bark of fleece that few, if any, shepherds would invest in having shorn. And once these lambs are dead and skinned and hanging in a cooler, their pelts—at New England abattoirs, anyway—are likely to be tossed into a foul-smelling dumpster with assorted "inedibles"; at this point, it's extremely unlikely that a lamb's wool will ever be shorn. But wait, the shepherds' lobby somehow argued with conviction. We did grow that wool, and we ought to reap some benefit. Wool, after all, is a strategic commodity. So for a time, when I took my lambs to slaughter, the abattoir's bookkeeper would verify their unshorn state. Forms would be filled out and dutifully filed; some months later I would get a check from the government for wool that had been growing on my dead lambs' backs. It was too good to be true, but there it was. Free money.

All good things do come to an end, however. Along about 1991, it was discovered that the world price of wool had been propped up artificially for many years by a rival wool program set up for Australia's shepherds. Those guys have a *lot* of sheep, compared to which the U.S. is a very minor player. The Aussies' program held a certain quantity of wool off the market—a substantial quantity, as things turned out—in order to artificially prop up the price of whatever wool was actually being sold. The excess wool had long been set aside in warehouses; wool, it must be noted, is virtually imperishable as long as it's kept dry. By 1990, nearly five million bales of wool had been sequestered from the marts of trade and stockpiled in Australia. A bale of wool has no fixed weight, but a good working figure would be three hundred pounds. Seven hundred fifty thousand *tons* of wool, then. That's a lot of uniforms. After the breakup of the Soviet Union—a longtime importer of Australian wool—the cost of keeping all this product held off the market became too much for even the Australians to bear. The warehouse doors were opened and the product went out; the world price of wool dropped dramatically, overnight. Dropped like a stone.

The ripple effect of this catastrophe didn't take long to reach Vermont, and when it did, hapless shepherds like myself found themselves being offered ten cents a pound for their crop of wool—or well under a dollar for a typical fleece. This was less than what you had to pay someone to shear it. Suddenly it seemed quite silly for our government to shell out all that money to ensure a captive wool supply. If our armed forces needed wool, they could buy it on the cheap from Down Under. Or from just about anywhere else. So the U.S. Wool Act went into a phase-out, and in 1996 the last checks went out. Our modest flock of about one hundred ewes became a one-crop enterprise—based on raising slaughter lambs—rather than a two-fer. But there was a takeaway that I made mental note of: from time to time, I realized, the government will spend its hard-earned dollars on the strangest things. Getting a piece of that ever-shifting action depended on knowing what particular things were being funded at a given time—that, and having the good luck to discover you were eligible for a program being offered.

Qualifying to receive a yearly "wool check" had not involved much personal interaction with the feds; once a year I took my collected receipts to the county agricultural office, and a schoolmarm-like woman would go over them with care. Certain *t*'s would have to be crossed, certain *i*'s dotted, but I could be out of there in ten or twenty minutes. By the 1980s, though, I came to be involved in a more substantial partnership with the U.S. government's bureaucracy for farmers. First, Cheryl and I thought it might be nice to build a pond near the central bowl of our premises, damming up the lazy stream that drains much of the valley. A good-sized pond, too. Maybe like an acre. This would not only enhance the landscape's scenic assets, but would give our growing family—with two small children, now, Ethan and Anaïs—a place to fish and skate and swim and just hang out. A private sort of Walden, to accompany our handmade lives. What better way to express our chosen values?

I bought a book on pond construction and learned that the government—doing business, this time, as the U. S. Soil Conservation Service—liked to see farm ponds built and liked even more to engineer them for a farmer. *Gratis*. Why would they do that? For one thing, ponds were a benefit to wildlife. And in a drought year they could be a source of irrigation. But when and if a pond's earthen dam became breached, valuable topsoil would be quickly swept away. Sometimes quite a lot of it, in an event that could take out road culverts and turn downstream lakes and rivers chocolate brown. Government engineers were trained to design dams that could withstand an estimated hundred-year flood—a flat-out cataclysm—whereas your average bulldozer jockey only worked to standards of by-gosh-and-by-gum. The author of this book on ponds urged that I play it safe and check in with the government before having a dam built.

So I paid a visit to our county's Soil Conservation office, and they sent a pair of experts out to see our pond site. Sure enough, their people were eager to participate—or I think the buzzword of the moment was co-operate. I was a co-operator, too. It was like a party. Survey stakes were driven in the ground, and numbers scribbled on them. One man scoped the landscape's contours with a transit level, and from time to time I'd get to hold a calibrated rod. It was an exceptional site, said the engineers. And because the watershed it drained was substantial, our pond would require an exceptional dam. They went back to their offices, drew up maps and worked out elevations; some weeks later they presented me with plans. The earthen dam was to be protected by an L-shaped culvert—eighty feet long, and thirty inches in diameter—to pick up normal overflow and whisk it safely downstream. And for major storms there was a broad spillway carved into the dam's southwest terminus, capable of clearing large quantities of water if the pond ever rose high enough to make this necessary. It was, I would have to say, an impressive set of plans. And I guessed that making them come true would be expensive.

When I showed the blueprints to a couple of local excavators—by now I had made the acquaintance of a few—they seemed astonished by the dam's proposed size and scale. Sixty feet across at the base, two hundred feet in length; altogether, bulldozing this work of earthen art would mean moving close to three thousand cubic yards of dirt. Or, in our case, heavy clay. Four thousand tons, roughly. Few had a bulldozer big enough to tackle that, or to do the job with appropriate efficiency. But for that matter, none thought that a dam so substantial and expensive was absolutely needed. Once you start playing ball with the government, though, you have to play by the government's rules. Finally I found a man who not only had a giant dozer in his arsenal of earthmoving equipment, but who also had a hankering to get into the sheep business. I say "business" tongue-in-cheek. We had a long negotiation, going back and forth; finally I swapped him fifty lambs from that year's crop for fifty hours of digging with his twelve-ton machine. After that, the flag was going to drop on his meter—and indeed it did, so that some cash traded hands as well. Still, it was the best deal that I ever made with sheep.

But now the burden was on him to build a dam that would meet the government's standards. Soil Conservation engineers were out inspecting the site every couple days, badgering my man as if he didn't understand the job and finding ever more creative ways to tick him off. The operator was in my employ, nominally—which is to say that I was paying for his labor. Paying by the hour for his bulldozer, too. But in another and equally important way, he was working for the men from the government. If they liked his finished pond, other jobs of comparable size might come his way. The county's Soil Conservation team had lots of work for bulldozers. What was going on between them struck me as a hazing. One day, I remember, he was ordered to dig up the stump of what had been a giant tree—a wineglass elm that had died of Dutch elm disease. I had felled the stem and cut up most of it for firewood, but of course there was a fat stump emerging from the tree's vast root system. He was told to dig it up, *roots and all*, and bury the debris someplace downstream from the pond construction. Otherwise, the wood was bound to gradually rot in place, making an eventual weak spot in the earth. As soon as the engineers had climbed into their truck and left, my pond builder gave me a sly grin and rolled his eyes. "That stump? I believe I know where that one's going." And he proceeded to entomb it in the rising dam—just about the last place one would want it to decompose. Next time the engineers stopped by to check on things, they didn't mention that stump. They seemed to have forgotten.

So it went, in a game of wary cat-and-mouse. The men from the government did not understand how to operate a bulldozer—let alone come up with the four-figure payments required to keep owning one, month after month—but they had a genial contempt for those who did. The man on the dozer had a good idea of what he thought a hundred-year flood would look like, and it didn't call for a fraction of the overkill built into our pond's design. Two weeks later, after battling wet weather and nearly losing his machine—twice—in a sea of muck, my man had the government's dam completed to their specs. Cosmetically, at any rate. Everyone seemed happy enough with the results. But I had an inkling that I might be just as happy if I'd dealt with the contractor one-on-one, like grown-ups. And that our farm pond might have cost a whole lot less.

Once the pond had filled with water, suddenly we began to see bats at night. After an evening's cookout on the grassy shore, we could see them taking to the skies just as dusk moved in. Hunting, I supposed. Sometimes they would dive-bomb the surface like kamikazes, then fly straight up and dance off toward the woods. No matter how keen your vision, it was hard to keep your eyes on one for more than just a couple seconds. They had these herky-jerky, skittering maneuvers. Now you see it, now you don't. Our kids would be fascinated, but we told them that once the bats had started coming out it was a sign for us to head up to the house. Humans and bats, we told them, don't really mix. For us, at least, the pond party was over for the night.

And then came the night when Cheryl and I awoke to hear a bat flapping above our heads, right in our own bedroom. We had a mutual panic attack. I switched the light on as we fled from the room; then I cracked the door and watched this jet-black specter circling the room at unbelievable speeds. Our bedroom is square, and yet the bat could turn square corners. Curtains hang down over the windows, but it dodged those, too. Finally it settled on the bed, perhaps exhausted. "How did it get in?" asked Cheryl.

"I have no idea. But how do we get it out?"

Cheryl brought a towel from the bathroom and I managed to open it wide and throw it on the resting bat, trapping it beneath a broad expanse of terrycloth. That afforded time to think, but it offered no solution. Now we had a bat beneath a bath towel on our bed. No good. I knew better than to touch a live bat and risk coming down with rabies. Could I maybe scrunch it up inside the towel and toss it out? "Let's open a window," Cheryl said after a moment's thought. "Then yank the towel away." That made sense to me—and then we turned the light off, too, since we knew that bats were used to flying in the dark. We ducked out again just as the bat took off; almost immediately it found the exit. After shutting the window good and tight, we crawled underneath the sheets with pounding hearts. Putting in a farm pond had brought us many benefits, but it had also brought bats into our lives.

The pond-building episode might have taught me to avoid joint ventures with the government—not because of bats, but on account of seeing how the government liked to do things. Liked to ride herd on folks and tell them what to do. Rather than learning the appropriate lesson, though, I went the other way and signed on for three years of having government diversion ditches built across the farm. Nearly a couple thousand feet of ditch per year, until the land would gently drain itself after spring runoff or even a hurricane. The trough of each ditch had an engineered slope of just one-half of one percent, or virtually nil; this required hundreds of precisely marked survey stakes and careful attention to detail once the digging started. But I could appreciate the benefits of all this work. I had seen neatly terraced hillsides for Italian vineyards; I had seen the burying of drainage tiles in Illinois cornfields. I had seen photographs of flooded Asian rice paddies, for which the engineering must have been no simple feat. If you're really serious about growing food, though, it pays to make investments in the land you plan to grow it on. In our situation, the fields thus protected would not only be spared any serious erosion, but they would be hay-able much sooner in the year. Taking off a first cut by the end of May, perhaps, rather than waiting till the fourth of July. That would allow an extra cutting per year—and the cuttings would occur at times when the crop was at its peak nutritive value, rather than grown stemmy and full of seed heads.

Of course, these ditches—like our beautiful new pond—were designed by the government at no real cost to me. And this time, the government was even helping pay to dig them. They were in for up to so much per running foot, depending on the final bill; I was in for however much would be left over. This put a premium on finding an operator who thought he could do the job at not too much expense to me. A typical charge for hiring a bulldozer was, in the early eighties, fifty bucks an hour plus fuel for the machine. The price of diesel could tack on another hundred bucks per day. So once the government's share of the budget for a ditch was fully spent, I'd be in for like five hundred dollars a day—and this at a time when I was earning next to nothing.

By law and as a matter of simple fairness, the Soil Conservation Service could not recommend any particular excavator to its various "co-operators." They handed out a list of prospective worthies, and it was up to me to contact one and make a deal. Several dozer jockeys sounded less than keen on doing government work, and they warned me that their engineers were hard to please. "If you want a ditch, I'll dig a ditch," they told me in effect. "But I want to be allowed to do it my way." That was not an option, I explained—not if I wanted to tap government funds. And I did. But as the three-year ditch-digging project gradually progressed, I wound up with a different dozer operator every year. The first two got their fill of government agents telling them what to do and calling them out over every imperfection. Then, in year three, the man I hired for the job seemed to be in bed with the local Soil Conservation Service team. I say this because one day I stopped by the county office and everybody there was wearing a baseball cap with this particular excavator's logo. That may not have been a recommendation, legally—but it sent a message.

After three years, though, I was fed up with co-operating. There were a host of minor, petty annoyances; what really turned me sour, though, was that people from the government would show up anytime they liked and stomp around the farm as if they owned the place. They would bring in visitors to show off the work in progress. They'd park their cars and pickups anywhere they wanted to, not even stopping at the house to say hello. Then, for one reason or another, they'd complain to me—as if I were their lackey, or a lazy hired man. A certain stretch of back slope had been seeded down to conservation grasses as per contract, but the seed had not come up—when would I reseed it? Another stretch of ditch needed mulching with a ton of hay—what was keeping me from getting on that? One ditch terminated in a special headwall that I'd built out of concrete blocks to the engineers' design specs, but I hadn't yet installed a sheet-metal flap that they wanted to prevent mice from crawling up a pipe. Why not? In short, these Soil Conservation men were not much fun. And I had a lifelong aversion—as I've tried to show—against being told by other people what to do. Older men, especially. Some of the soil engineers were likely grandfathers; I was still getting used to having turned thirty.

All in all, I wasn't quite an ingrate to the Soil Conservation Service for the good things it had done for me. But I could see that our relationship had no bright future. They had other practices in mind for me—a lot of them. They wanted to help me get my barnyard paved in concrete; they wanted to run some drainage pipe around the barn. There were certain gullies that they hoped to help me fill. That said, I was just a small fish in their pond of clients. Out by Lake Champlain—where the land is pretty flat to start with—hundred-acre fields were being dead-leveled with earthmoving machinery to eliminate runoff. Every dairy farm that was still in business seemed to be getting a government manure lagoon; that way, farmers wouldn't have to spread manure on frozen ground for several months each year. Manure spread on frozen ground winds up in Lake Champlain, polluting it with nutrients that set the stage for algal blooms. But these lagoons were running fifty thousand bucks a pop. Lots of work for excavators. It was as if you couldn't farm without a bulldozer, and a lot of dozers were on hire by the government. But I grew determined, now, to go it on my own. And to get the soil engineers out of my life.

There was something else, too: we were running out of money to do soil conservation work, as well as many other things. Our salad days were over, and the Porsche was long gone. True, we owned a lovely farm—and I had built a warm and comfortable house for us. We were not in debt; we had a reasonable line of farm machinery and a healthy, highly productive flock of sheep. But we were not turning a serious profit, and we'd badly underestimated what it might cost to raise two children in the way that we'd discovered we wanted to. They were both attending a private elementary school with a countercultural, free-spirited philosophy—not because the local public schools were unacceptable, but because it seemed the nicest gift that we could give them. Both kids had after-school activities to attend; there was the inevitable driving back and forth. These were not the suburbs, so we burned a lot of gas.

In the misty past when we had moved to Vermont, our oft-stated goal was to live self-sufficiently. Most of life's necessities we wanted to produce ourselves; for goods that we could not produce, we hoped to do some bartering. And indeed we did heat with wood cut in our forests here. We ate mostly homegrown food, and even wore sweaters that Cheryl had knitted out of yarn made from our own wool. But there were a lot of interactions with the outside world that we found could only be transacted with money. When your child has a fever and you take him to the doctor, it's unlikely you can pay the man in lamb chops. When your pickup needs a clutch, the man who does the job will not take cordwood for his efforts. And when the town's yearly tax bill comes around, the Clerk expects payment in coin of the realm.

Not long after having moved to Vermont, my wife began directing a day-care facility in the town of Middlebury—seven miles south of here. That made a nice fit for when our kids were day-care age, and it brought in a much-needed salary. I kept trying to stay active as a novelist—particularly during winter months, when life retreats indoors—but after my outrageous youthful success, I found that making serious money writing fiction is awfully hard to do. My first ship had come in with flying colors; subsequent ships had been launched but gotten lost at sea. There was magazine work, and I signed on for a lot of it—more than a serious writer maybe should have. Often I was able to please editors, but sometimes not—even after undertaking two or three revisions. Often they would pay me when they said they would, but sometimes not. Sometimes they would publish what I wrote but send no check at all, having gone bankrupt in the intervening period. Compared to the vagaries of managing a farm—where you take big chances on the weather, prices, crops, disease—the vagaries of writing made the farm look like a sure thing. Looking at the years ahead, we seemed to be at risk of slowly drifting into poverty.

So in 1986 I started teaching workshop courses at the local liberal arts college—Middlebury—on a part-time basis. At first these were exclusively creative writing courses, where I had some real-world knowledge to impart. But the job expanded into screenwriting and film studies, then first-year seminars on various topics; after several years I had attained a fulltime status, even teaching courses in environmental literature. Emerson, Thoreau, and Muir. Bradford. Bartram. Crèvecœur. Authors that you'd think it takes a PhD to explicate. I was a hardworking, dedicated teacher—so much so that my colleagues came to overlook the fact that I had never gone to graduate school and was thus *ipso facto* unqualified to hold my job. Of course, it didn't hurt that I was able to express myself in words like *ipso* and *facto* when the need arose. Once I realized that I was going to get a monthly paycheck, though—for a predetermined sum, direct deposited—I vowed that I would never misbehave in such a way that it could cost me that job. Adept as I was at Questioning Authority, now I was determined to keep quiet and suck it up. Suddenly we found that we were able to live decorously, and even save some money. Suddenly our house had a computer, and our kids could ski. Suddenly we saw how we could pay for them to go to college. Self-sufficient living was no longer a priority, or even a desideratum; we had joined the middle class—and probably then some.

My wife, too, found herself taking on an ever larger role in off-the-farm employment. She became an architect of social-welfare programs aimed at families raising children who were feared to be "at risk"; usually the risk involved the fact that these kids' parents were still children themselves. Or in their teens, at least. Eventually Cheryl took a job with state government, and wound up in the cabinet of Governor Howard Dean—helping to fulfill the *Whole Earth Catalog*'s prediction that our restless generation would take over Vermont. Former hippie that she was, she often went barefoot in the corridors of power; for photo ops and formal occasions, she kept dress shoes in the governor's closet. But Cheryl's job entailed driving "over the mountain" to Montpelier more days than not; that was a 100-mile roundtrip, no small feat in summertime and hellish in the winter. We now had Cheryl's parents living on the farm in a stand-alone dwelling just down the driveway—their retirement house—and that made the childrearing/homemaking work go smoothly despite our daily absences to go to work. As for the flock of sheep, we set up a management cycle such that most of the labor took place between the end of spring and the end of summer; by the time the fall semester rolled around, there would be no lambs at all left on the farm—just the breeding ewes and a lucky ram or two. We learned to postpone breeding till around Thanksgiving, so that lambs would not arrive until the following April when the weather—most years, anyway— had gotten mild. And when our "late" lambing season did begin, I found ways to tap eager students from the college to help out in the barn. Often they would come to spend the night and learn obstetrics.

So in a way, it all worked out for us. But in another way, year by year we seemed less rooted on the property whose purchase had defined a youthful version of ourselves; we lived more and more in the offices and conference rooms and lecture halls that marked our days. The farm began to go downhill in subtle but important ways. There was still enough time during summer months to put up hay, so our open fields were not growing up to weeds and brush. The forests, though, were languishing. In my late twenties, I had harvested red oaks for the framing of our sheep barn; I had helped out in the mill where those logs were sawn. We had taken time to build some footpaths through the nearest woods, so that we could walk in them or snowshoe in the winter—so we could enjoy them, after all, and feel at home there. We had made a careful, thorough census of our sugar maples. And for the first ten years or so of rural life, our only source of winter fuel was firewood that Cheryl and I harvested together, working with a chainsaw and a sledge hammer and wedges. That was the kind of work I could no longer justify once I had a teaching job; within a year or so of entering the classroom, we had a furnace installed and started burning oil. Good thing, too, because no one was at home on winter days to put a log on.

Push really came to shove in January, 1998, with what we now call The Great Vermont Ice Storm. Three straight days of freezing rain and steady drizzle coated every surface in a two-inch bark of ice—every trunk, every branch, and every twig of every tree. Of course the power lines became coated as well, and they came down everywhere. At night, the dark and eerie quiet—no purring refrigerator, no hot air gently blowing from the registers—was punctuated by the occasional report of snapping branches in the distance. Snapping tree trunks, too. It was like a war zone. Each explosive burst of noise would echo off the back cliff, then bounce against the glass façade of our house. After the storm had passed and the ice had melted, I ventured into the woods to check for damage—and after fifty yards I sat down and cried. Carnage was everywhere—so much that it would take me years to get the woods cleaned up, even if that was all I had to do. But I had no time to spend on cleaning up the woods. After that, I made a conscious effort to avoid them; if I got too close, I felt reproached by their sorry state.

And that was why I didn't make a fuss some years later when Scott Darling, doing business as Vermont Fish and Wildlife's expert on Chiroptera, sought my permission to set mist nets in the woods here and try to catch some bats. Why not? They didn't feel like my woods, anymore. Whatever was happening in them—what trees were growing tall or getting sick and stunted, what furry creatures might be thriving there or going down—it was beyond my capacity to care about. I had other things to do: books to read, lectures to prepare, papers and tests to grade. Work that paid handsomely and that was deemed valuable, notwithstanding its level of abstraction. As for the fact that this genial bat man was working for the government—well, my wife was working for the government, too. The war in Vietnam was over; I had mellowed. On the night that Darling and his helpers came to do their thing, I wasn't even there. But my son was. And next day, he informed me that they'd caught some bats.

Not just any bats, either. Indiana bats. Not one, but two of them—each a pregnant female. Living, healthy specimens of *Myotis sodalis*, a species that had long been listed as endangered by the U.S. Fish and Wildlife Service. What did this mean, exactly? Well, it meant that I had better not propose a shopping mall, or a power plant, or a clear-cut of the forests here. But no such plans were on my to-do list. Since its first version was enacted into law in 1966, a major effect of the Endangered Species Act had been to limit—or prohibit—real estate development in areas where listed species had been found. And Indiana bats had made the Class of '67 list—the first compilation of species thought to be endangered. (All seventy-eight of those species were vertebrates, consistent with that early and naïve version of the law.) So after forty years of legal endangerment, a wildlife biologist who understood bat behavior—who could think like a bat—was able to catch specimens of *Myotis sodalis* with no real difficulty. On the first try, so to speak. GPS coordinates for where Darling had trapped his bats were entered in a database of similar finds; it seemed to me that Indiana bats might not be nearly as endangered as was believed. Not around here, at least. And indeed among the ranks of Darling and his counterparts across the eastern United States, there was an informal contest each year to see who could trap the first Indiana bat. In 2006, Scott Darling must have won.

What was particularly odd about his triumph, though, was *where* on the farm Darling had managed to catch them. Sometime in the 1990s we had been approached by a delegate from VAST, an organization whose acronym derives from Vermont Association of Snow Travelers. VAST is a large, effective snowmobile club that marks and grooms and maintains thousands of miles of trails across the state. Four thousand five hundred miles of trail, altogether. In comparison, the state's entire system of paved roads comprises just six thousand miles. VAST provides its members with proprietary trail maps that they can use to travel just about anywhere a car could take them, and many places that a car could not. They have their own suspension bridges crossing creeks and rivers. And their thirty-five thousand dues-paying members have a keen sense of discipline. They won't start riding till the snow has reached a certain depth, they won't go freestyling off the marked trails, and they won't climb on all-terrain vehicles—or ATVs—after the snow is gone and run them on the VAST trails.

Eighty percent of the VAST trails are laid out on privately owned land, by formal agreement with a host of willing landowners. Landowners agree to this because it means that snowmobile traffic will be channeled and controlled, and because the contract gives them legal protection in case some yahoo gets into an accident. What the local VAST club wanted from us was to run a trail for about half a mile through the forests and meadows near our farm's southern boundary, hooking up with other trails in the near vicinity. It seemed workable, since the suggested path was reasonably distant from our house and barns and animals. Snowmobiles not only make a fair amount of noise; the hardy breed who ride them are surprisingly nocturnal, tending to be out at all hours of the night. And so the farther away from us, the better. A trail to accommodate the VAST club's needs was cut through a stretch of woods—a very thick and scruffy stretch of woods, I would have to say, made up of low-value trees that had invaded a wet, abandoned pasture last grazed before our tenure here. Once off the trail, this precinct of the farm had a dark, depressing atmosphere; consequently, we would almost never tramp around there.

But that was where Scott Darling set his nets up and had proved successful. He had a knack for this; he had a particular scenario in mind. He imagined bats would use the VAST trail as a flyway, heading to the farm pond for a drink just as night began. They would be parched after roosting in a tree all day. Bats are mammals like ourselves; their young are born alive, and females lactate little drops of bat milk from their bat-size teats. That would make them good and thirsty after a long, hot day. Once they'd finished tanking up, they'd hit the skies and hunt for bugs. A single bat can catch and eat thousands of insects in the course of a summer's night. Enough to equal half of the bat's own body weight.

The pond, then, had become a magnet for attracting bats. And many other kinds of wildlife as well. Give a wild animal the habitat that it prefers, and you can expect it to show up sooner or later; over the years since digging the pond we had seen it draw in muskrats, snapping turtles, even otters. Fish arrived from somewhere—as eggs on the feet of birds?—and made themselves at home here. Then came a great blue heron to dine on them. Ducks began building nests; Canada geese began stopping by each spring and fall. The pond became a rest stop on their migratory journeys.

Thinking like a bat, there was more to the pond than just an easy drink of water. There was much to credit in the pond's vicinity. There was lots of "edge effect," where one sort of ecosystem bumps up against the next. Forest versus pasture, say. Water versus hay fields. There were canopy considerations to admire, too—the way the tallest trees in the forest occupy their space. Bats in flight are constantly at risk of being preyed upon. By owls, in particular—those other master pilots of the nighttime sky. The right sort of canopy offers bats protection. All in all, we seemed to have a good site for bats to feed in. That was why they had showed up here; that was why Scott Darling's night had been a great success.

One of the texts that I was teaching every year back then—in the environmental literature course—was Aldo Leopold's *A Sand County Almanac*. In it, he writes with considerable eloquence about our human capacity to *care* for nature (or, as he would put it, the "biotic community") in ways that no other species is able to. We have caused much harm and destruction, to be sure. But we can also be protectors if we choose to be, since we possess both keen intelligence and a capacity to understand the consequences of our actions. And we also have the unique ability to grieve over what's been lost. Now—after the bat man had come here and done his thing—I had an anecdote to offer as enhancement to my lecture notes on Leopold. Were the human species endangered, I would tell my students, we could scarcely count on the bats to help us. But we humans have recognized that these Indiana bats are threatened, and in our human way we've set out to—what, exactly? To keep tabs on them. To monitor their presence, insofar as we can do that with our scientific tools. Mist nets. Miniscule transmitters and featherweight antennae. Telemetric tracking gear. To check on how the bats are doing, and to wish them well. That's an extraordinary thing, I would say with a measure of conviction. But equally extraordinary—and fairly troubling, too—was that this near encounter with a listed species had become a mere digression from a set of lecture notes. A bit of color to enhance my exegesis of a text. What had happened to the young man who had left Los Angeles and relocated to Vermont, seeking a direct—and a Thoreauvian—engagement with the world around him? When had I stopped engaging?

Meantime, in the world around me bats were about to take a serious nosedive—and not just as a figure of speech. In the winter of 2006–07, Scott Darling's counterpart in New York State—Allan Hicks, working for their Department of Environmental Conservation—began distributing a now-famous photograph of hibernating little brown bats (*Myotis lucifugus*) huddling in a cave not far from Albany. Actually it was in a portion of Howe Caverns, which has been a tourist attraction since the 1840s. The bats' noses were dusted with what looked to be a whitish substance, something like powdered sugar. Experts on Chiroptera routinely visit known hibernacula in winter months to conduct a bat census; bats can live for many years and will return to the same cave again and again. Compared to catching them in mist nets during summer months, finding bats and counting them in caves as they hibernate is more of a sure thing. But bats with powdered noses had not been observed before, Allan Hicks concluded after querying his colleagues. And there was something else: the cave's floor was littered with the carcasses of bats who had expired in there. Other caves near Albany were full of victims, too.

A year later, Allan Hicks phoned Scott Darling on a wintry night—at home—to let him know that a spelunker had just sent him a photograph of bats in a Vermont cave that was a known hibernaculum. There are quite a few of these near the southern end of the state's famous marble belt, near Manchester and Danby; several are not caves *per se* but mines where marble has been quarried. These particular bats, discovered in Mt. Tabor, had the same white powdery substance on their noses. That set off a wave of organized surveillance that confirmed that "white-nose syndrome" was now affecting bat populations in Vermont. Often the infected caves were rank with the odor of decomposing bodies, repelling even the hardiest investigators. Sometimes, though, the bats would leave their caves in midwinter and expire on the driveways and lawns of nearby houses. Sometimes they would make it through the hibernation period (November to mid-April, roughly) only to emerge in a weakened and necrotic state. By no means were all the known bat caves infected, but the mortality rates in those that had been were alarmingly high. Eighty to ninety percent, by some estimates.

Environmental journalists were quick to get wind of the story and jump on it. Doomsday articles began appearing in *The New York Times, The Boston Globe, The New Yorker,* and other venues. On TV, reporters were seen conducting interviews standing at the snow-covered entrances to bat caves. The usual conclusion was that once again the ecosystem had gotten screwed up—"deranged"—and that in all likelihood humans were to blame. All of this attention might have backfired on the bats, however. One hypothesis was that spelunkers had unwittingly spread white-nose syndrome from site to site. If each bat returns in the fall to a particular hibernaculum, and if white-nose syndrome thrived only in cave environments, how was it getting distributed from cave to cave? Quite possibly by recreational cavers. But if all this blitz of media attention brought even more people into known bat caves—to see the problem firsthand—then the situation was likely to get worse.

And indeed it did. One particular cave in Dorset, Vermont—Aeolus Cave, named for the Greek god of the winds—had had its winter bat population studied annually since the 1930s. It's an exceptionally large and deep hibernaculum, thought to have been used by bats for ten thousand years. In 2008, Darling reported that the stench of dead bats near the cave entrance made it impossible to enter and conduct surveillance. In 2009, he estimated the total mortality in that one cave to be at least one hundred thousand. And of thirty known hibernacula around the state, only four at that point showed no signs of white-nose syndrome. Two of the state's nine bat species were deemed to be at particular risk: the little brown bat (*Myotis lucifugus*) and the northern long-eared bat (*Myotis septentrionalis*). The former had at one time been the most common bat in Vermont, with an estimated population of six hundred thousand—or about one such bat per human resident. Now they were becoming the state's rarest bat species.

Such a quick decline for what had been a robust species doing business as usual in its longtime habitat was hard to comprehend. In most disease scenarios, there will emerge a cohort of the target population that possesses some genetic advantage that allows them to withstand the biologic challenge. That idea is central to how evolution works: those who survive become improvers of the breed as they pass on their superior genes—and their resistance—to successive rounds of offspring. But in the case of bats, rebuilding a decimated population would take time. Only half of bats, roughly speaking, will be females. Females bear only a single pup per year. Half of those pups can be expected to be males. Of the females, some will surely die before they procreate; others may grow to maturity but prove infertile. In the absence of any direct way to deal with the cause (or the causes) of white-nose syndrome, all that wildlife biologists could do was hope some remnant of the decimated species would hang in there—and do so long enough to gradually rebuild their ranks. That would require more than waiting and hoping as the bat population got winnowed to a lucky few. On the other side of that winnowing process, bringing back the bats would mean providing them with habitat.

Forest Management Plan

In 2008, I took a one-semester leave from my teaching job and spent some time giving thought to my next move. I was sixty-one years old—getting kind of creaky, but still able to do some hard physical work. Our children had both turned into interesting adults, both in strong marriages and both more than able to take care of themselves. A rising generation of Young Turks was attaining tenure at the college; as Young Turks in higher education tend to do, they were "reforming" the curriculum—not because it was in desperate need of reformation, but as a way to knock heads with their elders now that they could not be fired. As a mere lecturer, I had never been required to squeeze through the tenure gate, so I was not one of the elders being challenged. But I was generally aligned with their entrenched perspective. In order to cobble together a full-time job, I had gotten teaching work in three different departments—and each of them seemed to have become a sandbox where PhDs set traps for each other. Department meetings were like tournaments of chess.

I had certain other perceptions to acknowledge, too. The gap between my age and that of my students had grown wide, and would grow ever wider. Several of my courses dealt with popular culture—film, in particular—and every year the list of films that I considered worthy of critical attention seemed more eccentric in my youthful students' eyes. I saw older colleagues, too—more than a few of them—who had lacked the wit or courage to hang up their skates when it might have been smart to. Now they had become too old to reinvent themselves, and they seemed ensnared by their status perks, their privileges. I knew that I didn't want to turn out like one of them. And I had a yearning to spend time on the farm again—to feel like I actually lived on the premises, rather than being an absentee landlord. I wanted to reassert the way I thought the place *should* look, rather than what forces of entropy had dealt it. True, I'd miss the paycheck if I quit my fancy teaching job. And the many benefits—Middlebury College was a generous employer. But I'd saved some money up and the time felt right. I gave notice to my relevant superiors, and a year later—in June, 2009—I set myself free.

An immediate priority was trying to reduce expenses, and property taxes on the farm were a major item. When we had acquired the place, the Town of New Haven's tax bill was around eight hundred dollars a year; over the intervening decades, it had grown tenfold. Of course, I had done many things to raise the town's assessment. I had built a spacious and comfortable house—with an indoor hot tub and a basement sauna—as well as barns for animals and hay and machinery. There were two gazebos I had built astride the farm pond, joined by a slate terrace; one was mainly used for preparing meals on summer nights, and the other was where we consumed them with friends and family. Cheryl had a meditation hut—a *poustinia*—nestled by a natural amphitheater in the woods. I had no serious cause to protest our taxes, all things considered. But I did badly want to see them get reduced.

In Vermont, a landowner can usually accomplish this by enrolling in the state's Use Value Appraisal Program. UVA, for short. Or, in common parlance, Current Use. Ordinarily, a property's appraised value is tied to what the town's listers feel it could be sold for—tied to its "fair market value," that is to say. Under Use Value Appraisal, though, the valuation is adjusted to reflect how the land is actually being used. Pastures for grazing sheep, in our case. Meadows for making hay. And many acres of forested land that were suffering from neglect. Thanks to this program of property tax relief, a lot of farms like our own had been enabled to continue in agriculture—even when it was only marginally profitable—rather than being sold for development.

I had heard of Use Value Appraisal for many years; our town included basic information on the program with each year's tax bill. Neighbors had advised me I could save some dough by signing up. But there was a catch that I was keenly aware of: before someone can place a *forest* parcel in the program, they have to present a Forest Management Plan. Drawn up by a recognized forestry professional. Then the county forester—and every county had one, employed by the state's Agency of Natural Resources—would have to review the plan and hopefully approve it. Then the county forester's job required making sure the plan was carried out. So many thousand board feet of lumber to be harvested in 2015, for example. So many cords of firewood to be extracted. So many acres to be thinned by a certain date. During all my years of teaching, I knew that I had no time to spend on such activities—and I knew I didn't want to hire someone else to do them. Since I'd found my way into a job that was paying well, it made sense to suck it up and pay the tax bill. Participating in the Use Value Appraisal program might have saved us lots of money over the years, but participating in it would have cost me precious time. Simply stated, I had better things to do. It was a helpful piece of legislation, probably—but not a free lunch.

Nor did I believe it should be. I had read my share of environmental history; I knew that a nation's forests are a fundamental resource. Bad forest practices have changed the course of empires. In Spain, for example. And in Italy, and Greece. Even Great Britain—by the time of the American Revolution, the monarchy had become perilously dependent on trees from the New World. Trees were required to build ships, after all, and England was a naval power. All across her North American colonies, white pines of the size and girth that made them likely ship's masts had been marked with a "broad arrow" blaze to show they were the king's. They could not be cut unless his agents gave the word to do so, and then the tree was his. That stirred the pot of simmering resentment, leading to revolt and eventual revolution. But back at home in the British Isles, there was no comparable forest resource. The land had long been cleared for farming.

When forests are managed by a monarch and his minions, the blame for heedless management decisions can be laid to them. Where forest ownership is privatized and atomized—as is the case across much of New England—a lot of people have the right to make some bad decisions. But the consequences of those choices will affect us all, sooner or later. So I understood why the government wanted to have landowners come up with a Forest Management Plan—to be enforced by government agents, if need be—before giving those same landowners a tax break. I just didn't think the program made sense for me. I had little interest in allowing the state's bureaucracy of foresters to get on my back.

Now, though—after choosing early retirement—getting the farm enrolled in Use Value Appraisal seemed essential. It was bound to pare our expenses, and they needed paring. I had the time and will to get some work done in the woods, so keeping in step with a Management Plan wouldn't be impossible. I paid a visit to our county's agricultural office, which had moved to a new location since my last visit there twenty-five years before. The old office might have been designed as a cheap motel, even a cathouse; the new place was sleek and swank and set in a spiffy new industrial park. There were no familiar faces when I approached the counter; no one knew who I was, and I didn't know anyone. The county forester—a man named Chris Olson—was out on a site inspection, but I left a message and he called the next day. He wanted to meet at our place, which made sense to me—I expected he would want to walk around the woods. No, though; when he showed up, making a physical inspection was not the point. We sat down together—Olson, me, and Cheryl—at the dining room table, from which he could look out the window and see distant trees. He had already checked out our forests via computer, pulling up topological maps and images taken from low-flying airplanes as well as orbiting satellites. He had lots of information. What he'd come to do was just make sure we understood the program. Use Value Appraisal had some fine points, and some caveats. And I think he wanted to scope out whether Cheryl and I were apt to hold up our end, if he let us in the program.

On his office wall, he said, he had a county map with all the forest land enrolled in the tax program delineated. Land that was under his nominal control, inasmuch as he could kick people out of the program if they didn't stay current with their Management Plan. Many thousands of acres had been enrolled—a significant portion of the county's private forest lands. "But some of these people, years go by and I don't see them working in the woods. Finally I have to say, 'Listen—you've got one hand reaching out to grab a dollar, and the other's in your pocket. Do you think that's fair?'" I agreed it sounded like a terrible thing. But in the next breath, I asked Olson if committing to a Forest Management Plan would require that we harvest timber in a year when prices for stumpage had been beaten down. Like at the present moment—we were, after all, in the throes of a recession. Construction had tanked, so demand for logs was very weak. The county's biggest sawmill was converting some capacity to turning out *firewood*, dried in kilns that had been built for drying sawn lumber. I said that I didn't want to sell logs in a time like that, whether a piece of paper had declared I would or not. No problem, Olson told us. Forest Management Plans do specify a certain year for this or that harvest to be taken, but they give the owner three years of leeway from the targeted date. Three years in each direction. So a timber cut that had been scheduled for, say, 2015 could actually be carried out as early as 2012 or as late as 2018. That was some window, I had to admit. And a person wasn't likely to be spanked for noncompliance until ten years after getting in the program, when their entire Forest Management Plan would be up for review and possible renewal.

Then Olson told us we had plenty of latitude in choosing which acres of the farm to "put in" the tax relief program. It only made sense to "put in" acres that we had no thought of selling for at least ten years—since to ever take land *out* of the program would require paying the state a Land Use Change Tax. The tax rate for land sold within a decade of enrollment was twenty percent of the fair market value as determined by the state. After ten years, though, the rate dropped to ten percent—a flea-bite, if real estate continued to appreciate. We also weren't allowed to "put in" forest land that had already been developed—for example, that had an existing structure on it. But an existing structure could be overlooked if it was held to be pertinent to forestry. Some of his participants, he said, kept hunting camps hidden in some secret corner of their woods. Places to drink beer with their buddies during deer season. That was a no-no, from the state's point of view—not the drinking, but the structure's presence on land that was receiving tax relief. So he'd tell those hunters they should keep some tools in the cabin. Like an ax. Some files for sharpening a chainsaw's teeth. Or perhaps a splitting maul. Then the offending structure could be construed as involved in the day-to-day practice of forestry.

"How about a *poustinia*?" asked Cheryl.

"Come again?"

"A meditation hut," I told him. "Cheryl goes to pray there, in the morning. And to read the Bible." Also the Koran, I might have added. And Gilgamesh. And a small collection of other sacred and quasi-sacred texts.

"This is in the woods, you say?"

"Right along the edge of them."

"How big?"

"Six by ten."

"Six by ten *feet*?"

I nodded. "It's not all that big."

"Think there's room to keep a hatchet?"

Cheryl felt sure that there was room to hang a hatchet up, so her *poustinia* would not become a deal breaker. The main decision that we had to make, said Olson—and to make rather quickly—was choosing one professional forester or another to draw up our Forest Management Plan. This had to be presented in conformance with some guidelines that the state had handed down. Then, like the agent from the Soil Conservation Service thirty years ago—the one who'd handed me a list of earthmoving contractors, but wasn't legally allowed to recommend one—Olson now pulled out a list of consulting foresters. Just to have one come out and take a look around, he said, would cost around a hundred dollars. Fifty bucks an hour was the customary rate. But it might make sense for us to do this dance with several people, getting a feel for the kind of partner we could live with. Once committed to working with a forester, it was a longtime proposition. Almost like a marriage.

"So let's say we find someone we like," I said, "and hire him. How much are we talking to come up with a plan?"

"That would all depend," said the state's man predictably. But we would be asking for a fair amount of someone's time—going through the woods with deliberate care, making notes on what was there and taking lots of measurements. Then the different stands would have to be described in special language; forestry professionals had their own vocabulary. We'd be in for about a thousand dollars, easily. Maybe more like fifteen hundred. Then, looking down the road, we'd have to re-engage this person whenever the plan called for some action to be taken. Like a timber sale, say. Our consulting forester would want to mark which trees to cut, help us choose a logger to perform the operation, and then manage the sale of our timber to a sawmill.

I looked over the list of private foresters, and knew right away which one was my first pick. Vermont Family Forests had been started by a man I'd met on several occasions, over the years. David Brynn had actually had Chris Olson's job—Vermont State Forester for Addison County—for fifteen years before entering the private sector. He was known to be a New Age sort of guy; his focus was on fostering healthy forest ecosystems rather than merely the production of trees. I knew these things primarily because my son's wife, Susannah McCandless, had worked for Vermont Family Forests as a researcher during her doctoral studies in geography. Their office was just ten miles away, in Bristol. "How about this outfit?" I asked, pointing to the listing on Olson's printout of forestry consultants.

"They can do a good job, if they have time for you. Better call them right away, if that's who you want to go with." It turned out that we were bumping up against a deadline for new enrollees to get into the program. We were well into June; applications had to be received by September 1. Foresters tended to be overworked this time of year, writing plans for new clients or revising old ones. Vermont Family Forests was busy, Olson said—and maybe somewhat understaffed. We'd be foolish to sign up with someone who might miss the deadline, or do a hasty job to get things filed on time. Once a Forest Management Plan had been accepted, we would have to live with its directives for a decade. But if someone did turn in a slapdash plan, I wondered, wouldn't Olson pick that out and turn it down? At the same time, I was putting two and two together. If our Forest Management Plan had been drawn up by a former county forester, surely it should pass muster. Surely it would stand up to whatever specs the state demanded. I said I would find out whether David Brynn could take us on, and Olson left us with a sheaf of forms and information.

Next morning, I called Vermont Family Forests and went out to do farm chores; by the time I got back to the house, Brynn had returned my call. He already knew a fair amount about our forests here, since he had discussed them with my daughter-in-law. Yes, he'd love to spend a couple hours reconnoitering. And he wanted to bring an associate along with him—Brendan Weiner, a young man who had lately joined his staff and would be the person drawing up our plan if we went with them. For my part, I wanted to have Cheryl on the walk with us, and our son—Ethan Mitchell—and his wife, Susannah. Getting all six people on board took a bit of juggling, but eventually we came up with a time that everyone could make.

David Brynn is my age, more or less, and of average build—not the sort of person you'd mistake for Paul Bunyan. His blondish hair is thick and frizzy, something like Gene Wilder's. Or Art Garfunkel's. Brendan Weiner, on the other hand, was tall and broad-shouldered; he looked like he'd taken down a serious tree or two. He'd recently completed a master's program in natural resources at the state university; prior to that he had worked for the U.S. Forest Service out West, restoring damaged ecosystems and fighting wildfires. As the six of us tramped all around the woods that day, Brynn did most of the talking while Weiner ran a GPS device and snapped digital pictures. He didn't say much, but I saw that he was on the case. Gathering data points to put into our Forest Plan.

What we had to do, said Brynn, was think in really bold terms about what our goals were as owners of these private woods. Did we see them as a place for personal recreation? Did we want to hike in them?

Cross-country ski, or snowshoe? Did we aim to use them as a means of finding solitude? What about wildlife—might that be a priority? What about protecting the quality of water, or enhancing scenic beauty?

"Gee," I said. "I thought this forest program had to do with trees."

"That used to be its main thrust," admitted Brynn. "But this past spring, the legislature changed things." What he didn't tell me was that he himself—as well as several colleagues at Vermont Family Forests—had been instrumental in pushing certain changes through. Up until a very short time before the day we spoke, having a "timber focus" was what county foresters were supposed to look for when they judged a Forest Management Plan. In exchange for tax relief, the landowner would improve the quality of trees in the woodlands under management; that would guarantee the future of a job-intensive industry deemed to be important to the state's economy. But what Brynn and others had successfully argued—in a hundred-page report prepared for the lawmakers—was that under some conditions forest parcels ought to be allowed into the program even if there was no foreseeable timber harvest. For example, vernal pools were not productive forestland but they did serve other functions—such as giving frogs a place to breed in the spring. Forested wetlands, too—like our sprawling beaver swamp—were virtually useless as land for growing trees on, but they had other ecological benefits. There were many new and offbeat arguments that could be made for why certain parcels of land should be awarded tax relief. All that was required was to have a professional forester make the case, and make it in sufficient detail to be convincing. Bottom line: the door had just been opened for unusual, outside-the-box Forest Management Plans that would not likely have been approved in any prior year.

When my son was in high school, he and several friends built a trail looping through the woods that came out at a bare expanse of rock atop the back cliff. A private aerie, with sensational views toward Lake Champlain and the Adirondacks. Now we tried to find that trail, and we found it badly hidden under downed trees and a chaos of fallen limbs. It was more like bushwhacking than hiking on a beaten path; it would take a couple weeks of dedicated labor just to make this modest footpath passable again. I felt embarrassed by the rundown state of things. Brynn, on the other hand, kept admiring aspects of the site—deadfalls notwithstanding. He saw evidence of soil fertility, and he liked the south-facing aspect of the woods. He pointed out some particularly "happy" trees, able to grow with both efficiency and grace. He seemed to be seeing things that Cheryl and I were blind to.

Quizzing him, I learned a few basic tenets of professional forestry. A tree's size has no fixed relation to its age, despite our tendency to think of them as being connected; size has more to do with the site's capabilities than with how many years a tree has been growing there. A given piece of woodland—like an acre, say—has a certain growth potential in cubic feet of wood per year, and that potential is essentially fixed. What forestry can do is to direct that growth into trees of higher value than would otherwise be the case. Trees that would contribute to a healthier forest, too. There were certain well-tried strategies for doing this, but each involved trade-offs between rewards and risks. Natural disturbances—like The Great Ice Storm of 1998, evidence of which was still lying all around us—were a constant possibility. Even a certainty, if you took a long view. So in some ways, forestry would always be a crapshoot. Particularly if its only goal was to procure sawlogs. But if you worked to make a healthy forest ecosystem, valuable timber would inevitably grow.

At some point, Brendan stopped taking coordinates and joined the conversation. "You have a lot of shagbark hickories," he pointed out. We were passing right beneath one, and it looked impressive.

"I tried cutting one for firewood, once," I said. "Never again."

"Why not?"

"Too damn hard to cut. And then the wood won't split."

"You know who likes shagbark hickories? Bats."

"Do they?"

"Yes, they do. It's supposed to be their favorite roost tree."

Why on earth should that be? I wanted to know. Well, the way the bark tended to split off from the trunk and hang down in ragged shingles—the way it *exfoliated*, I think I heard him say—made for lots of nooks and crannies for the bats to crawl up into. They would take cover there as soon as dawn broke after a night of hunting bugs, then sleep through the day. A big shagbark hickory tree could be a bat hotel.

"Bats are endangered, by the way," commented David Brynn.

But of course I knew that. "Actually, someone from the state caught Indiana bats here a while back," I told them with a certain pride. "Three or four years ago, I think. Guy from Fish and Wildlife. Making some kind of survey."

"Scott Darling?"

"That sounds right." I recognized the name.

"That could be really useful, looking down the road."

That's about as far as the conversation went, that day. We did engage Vermont Family Forests to draw up our required Management Plan, and Brendan Weiner came back several times to cruise the woods and analyze them carefully. Seventy acres of eligible forest land; for management purposes, he broke them into four zones with separate directives. One of these zones had a couple of natural communities recognized as being "unique and/or fragile" by the state. Specifically, there was a *temperate calcareous outcrop*—that is to say, a cliff—and a *red cedar woodland* associated with that cliff. Preserving these communities pretty much gave us a pass in Zone #2, under the newly adopted program regs. Good thing, too, because that portion of the woods was extremely steep and hard to get to. Consequently there was no way—short of a helicopter—to get logs out.

What was really tantalizing, though—from the point of view of testing the state's new guidelines—were the shagbark hickories scattered through zones #1 and #3. True, they were growing nowhere near where Scott Darling had caught Indiana bats in the summer of 2006. But elsewhere in the woods they were a fairly common sight. Not just pole-sized specimens, either. Big, thick, towering shagbarks that had fought for a place in the forest canopy and sometimes had achieved a "dominant" status there. Or at least "co-dominant" status, nearly just as good; these are terms of forestry description I was learning now, and their underlying concepts weren't that hard to fathom. Dominant trees are the winners in a forest landscape, capturing the lion's share of sunlight and moisture and soil nutrients. Co-dominants are the runners up, so to speak. Since it had been documented that we did have bats here—and bats of a species that was federally endangered—and since we had a nice selection of the very tree that bats preferred to roost in, we could shape our Forest Management Plan around the goal of helping bats. Offering them habitat, and trying to enhance that habitat's quality. That would take the pressure off scheduling a timber sale, which would have meant losing some of our biggest and best-grown trees. I could spend my time getting the forest floor cleaned of debris and laying out trails; then we could enjoy the woods whenever we were able to. And get a break on our property taxes, thanks to the state's Use Value Appraisal program. All we'd have to do was offer trees for bats to roost in.

For reasons too complicated to go into here, Cheryl and I spent the end of that summer in South Korea; long story short, she was doing some consulting there and I was asked to come along. Meanwhile back in Vermont, though, the pressure was on to get the farm into Use Value Appraisal by the looming deadline. I recall a flurry of e-mails back and forth—with Brendan Weiner at first, and then later with a woman from the Wildlife Division of Vermont's Fish and Wildlife Department, which exists within the Agency of Natural Resources. Jane Lazorchak was coordinating something called the Landowner Incentive Program—LIP, for short—a conservation effort aimed at helping species deemed to be at-risk. Deemed by the state, that is—not the federal government. She'd been tipped off by Brendan that our woods had some likely habitat for bats, and that he was planning to build that circumstance into our Forest Management Plan. Did she want to stop by the farm and have a look around? Jane did, along with an associate named Kristen Brisee. The bat ladies, I dubbed them from an Internet café in Seoul. I couldn't be on hand to give them a tour, but I figured they'd be able to find the shagbark hickories. I said to come on out and make a site inspection, and they chose a date to take a walk around the woods.

When I got back to Vermont a week later, Brendan approached me with an unexpected proposition. Also a somewhat maddening proposition, and in certain ways preposterous. The Landowner Incentive Program had discretionary funds that they were willing to spend to help us do good things for bats. The deal was that the state would pay for three-quarters of whatever invoice Vermont Family Forests presented for our Management Plan, and Cheryl and I would only have to come up with the rest. Twenty-five cents on the dollar, that is to say. It looked like the bill was going to come in at around twelve hundred dollars, so getting nine hundred of it paid for by the government was an attractive offer. What the state wanted in return was that our Forest Management Plan would specify "conserving and enhancing Indiana bat habitat" as our primary objective in managing the forests here. That exact language had to go into the document. There could be—there *were*—many other objectives, such as "recreation" and "enhancing scenic beauty" and "protecting biological diversity." But these would have to be acknowledged in the document as secondary goals. Habitat for Indiana bats would be Number One.

I said I would think about that, then had a long conversation with Cheryl. "Look me in the eyes," I told her. "Do I look like somebody whose *primary objective* as a forest owner is helping Indiana bats?"

"No," she answered. "But you look like somebody who'd like a handout."

Primary Landowner Objective. On its face, it was a ludicrous declaration. Now I started railing about other declarations that people in authority had gotten me to make, down through the years. Polished verbal formulae that clearly went beyond my childish powers of comprehension—but that didn't mean I couldn't learn to say the words correctly. By the age of three or so, every bedtime tuck-in ended with a certain prayer:

Now I lay me down to sleep;
I pray the Lord my soul to keep.
If I should die before I wake,
I pray the Lord my soul to take.

This strikes me as an outrageous set of thoughts to send a child to sleep on. Fortunately, three-year-olds have no idea what they're saying. "Lord," "soul"—even "die"—are vague and unfamiliar concepts, made even stranger when presented in subjunctive mood. Then there was the Lord's Prayer, and the Pledge of Allegiance—credos that we had to recite before each day of school. True, you can get first graders to commit a certain set of words to memory—but it will take them many years to understand their meaning. What purpose did the teachers think was being served?

"But you're not a child anymore," Cheryl pointed out. "You aren't being asked to say things you don't understand."

"I don't like people putting words in my mouth, is all."

"Then don't take their money. Nine hundred dollars? Come on—that's not a fortune."

But it wasn't chump change, either, to a guy who had begun to miss his paycheck. I hemmed and hawed for several days, then sat down with Brendan. What, exactly, would it mean if I said yes to the bat ladies' proposal? Brendan said that first of all, it meant I'd better not start cutting shagbark hickories. They were all potential roost trees—that was what the forest here offered, from the state's perspective. Up to one hundred possible Hiltons. Then there might be opportunities to "daylight" some of them, or to "release" them from surrounding forest canopy. That meant opening the woods around the shagbarks in ways that would hopefully attract a few more bats.

"That's it?"

"Basically."

"Then I might as well sign up. I didn't want to cut the hickories, anyway."

But there were a few more items on his to-do list, if we were going to enhance the woods as habitat. "You could use more snags," he told me. "We like to see at least six of those per acre."

"Snags?"

"You know—snag trees."

Actually I didn't know, but I was about to learn. Snags are dead or dying trees that haven't fallen down yet, but are in a state of gradual decay. So they have a growing wealth of crevices and cavities. Various woodland creatures—including certain species of bats—will choose a snag to make their home in. Snags and forest wildlife are virtually synonymous. But this raised an obvious question, when I thought about it. "What am I supposed to do to get more snags?"

"Choose some living trees of poor quality, and girdle them."

Girdling a tree—this rang a distant bell. I had learned about this back in Boy Scouts, or in junior high. Every tree, no matter how enormous and well-grown it is, has an Achilles' heel: cut through the cambium layer just beneath the bark in a continuous circle, and a tree will die. You could do it with a hatchet, if you had the time; in principle, you could even do it with a penknife. Doing it with a chainsaw was like child's play. Girdling even big trees could be done in a flash, after which they would start morphing into snags. But what kind of forestry was this? I had to wonder. It seemed more like vandalism. Going through the woods and killing trees, but leave them standing? It was a matter of ecosystem health, said Weiner. In a healthy forest, a certain amount of decay is desirable. As a girdled tree begins to gradually rot in place, it offers food and shelter to other life forms in the woods. Insects, for example. Who get eaten by the birds—as well as foraging bats. What I had to do was see the forest as an ecosystem, rather than a bunch of trees. A rich, diverse community of co-existing organisms.

As far as my forest's ecosystem was concerned, there was another problem Brendan had noticed and wanted me to know about—because the government was bound to notice, too. There was a modest but growing infestation of a plant called buckthorn. I could not recall having ever heard of buckthorn; certainly it wasn't present when we'd bought the farm. Brendan said he'd tag a couple specimens for me to look at right along the forest's edge—where the plants would make a sort of beachhead to get established—with bright red flagging tape. And then he would send me the address of a website where I could discover why this plant was such bad news. Brendan had a calm demeanor; he was not suggesting that I push some kind of panic button. But he clearly wanted me to get myself informed.

Buckthorn, I learned after a web investigation, has become a serious "exotic" or "invasive" across much of the Northeast. It is well on the way to becoming our kudzu. It does not belong here; it came from northern Europe, and in New England forests it is too competitive for the native plants. There were two varieties, "common" and "glossy" buckthorn; my main problem seemed to be the glossy kind. The dark green leaves have a distinctive waxy sheen, and they emerge quite early in the growing season—depriving adjacent plants of scarce and precious sunlight. They also stay green quite late into the fall, after all the other forest plants have shed their leaves. This extended growing season helps the buckthorn come up fast. Buckthorn seeds come packed in dark red berries—purple, when fully ripe—that birds eat, spreading the seeds far and wide in their resulting poop. Buckthorn is cathartic so the birds come down with diarrhea, adding to the plant's seed dispersal ability. Newly sprouted buckthorn might be taken for a pretty weed, but it soon turns into a shrub and then a gnarly tree—up to twenty-five feet tall, if no one intervenes. The plants often build up impenetrable thickets. Where open fields have been abandoned by bankrupt farms, buckthorn is well on the way to taking over. And along considerable stretches of our forest's edge, buckthorn was present and had started to do its thing. Even in the forest's well-shaded interior, there were buckthorns doing well amid the oaks and sugar maples.

So I filed that news away for future reference: I had an exotic and invasive species on my hands. At least the wood of buckthorn trees was fairly hard, said Brendan. It was dense enough to throw off plenty of heat, so I should consider it as possible firewood. Firewood was weighing on my mind, at that point in the year. Part of my retirement expense-reduction scheme had been to turn down the furnace's thermostat and burn a lot more wood again; the price of oil had begun to flirt with four bucks a gallon, just like gasoline. The house that I had built before my teaching career began was tight and well-insulated, and it captured solar gain across a glass façade that was aimed directly south. Whenever the sun was shining on a wintry day, we were toasty warm inside without supplemental heat. But there were cloudy days, too, and long winter nights; we had gotten used to buying—and to burning—over eight hundred gallons of fuel oil per year. Maybe burning buckthorn was a viable alternative, getting this invasive species out of the forest and into the woodstove. Killing two birds, as it were, with one stone.

A couple weeks later Brendan sent us his completed Forest Management Plan, as required by the state before we could put the woods into Use Value Appraisal. It was thirty pages long, and bound like a corporation's annual report. It had color photographs taken on the premises, showing what the woods here actually looked like. There was a foldout map, with clear delineation of the four different management zones that we'd agreed upon. It was, all in all, an impressive document and easily worth the pittance we had paid for it, since the state's taxpayers had been kind enough to help us out. But it wasn't easy reading to a nonforester. In fact, a lot of it read like so much Greek. I wound up buying a layman's guide to forestry just to understand some of the unfamiliar terms. Going back and forth between this guidebook and our Forest Plan, the data that had been recorded gradually made sense.

One term that I already knew was *DBH*, which stands for a given tree's diameter at breast height. You measure this by wrapping a tape around the tree trunk, measuring its circumference in inches and dividing by pi. Trees big enough to view as serious sawlogs needed to have a DBH of fourteen inches, minimum. "Pole"-sized trees were those that measured from five to nine inches DBH, and saplings were smaller than that. The goal was to somehow get a quantitative handle on how much wood was likely to be in a tree—or in an acre of trees, or in a forest. DBH calculations didn't tell you everything, but they were a starting point.

I had seen, somewhere in the murky past, wooden "cruising sticks" that you held against a tree to get a fix on its diameter; printed on the back side was a table of numbers—a "log rule"—that converted DBH calculations into board feet. A board foot is a square foot of wood one inch thick, and that is the unit by which lumber is bought and sold. If you knew the merchantable height of a standing tree and you knew its DBH, a log rule gave you an estimate of board feet. But *board feet* was not a term used even once in our Forest Management Plan. Instead there was a concept called the forest's *basal area*—or its basal area in square feet per acre. This meant imagining that all the trees had been cut off at breast height—four and a half feet, nominally—and that the area of each stump had then been calculated. Pi times each stump's radius, squared. If you did that for all the trees in an acre and then added the results, you came out with a certain number of square feet. And it was apt to be a pretty low number, given that an acre is 43,560 square feet. Eighty square feet, say. Or eighty-five, or ninety—that was typically the portion of a forest acre occupied by the trunks of all its trees at breast height. To a professional, this said more about a forest stand's density than DBH or estimated board feet ever could.

Another unfamiliar term was each zone's *site index*. Based on the soil type, hydrology, slope, orientation, and other factors, a number was determined—or a range of numbers, sometimes—that predicted how tall a tree ought to grow on that site in fifty years. Or how tall it *did* grow in fifty years, if it happened to be older. The site index number depended, to some extent, on the tree species grown or what the forester planned to grow. Then there were terms like AGS and UGS and *cull*, used to characterize the quality of tree stems as potential sawtimber. AGS stood for Acceptable Growing Stock; UGS meant Unacceptable Growing Stock, which meant it was below a certain quality grade but still might be sold. Cull meant—well, I already knew that one from our shepherding activities. Once a year we'd go through the whole flock, sheep by sheep, to cull the ones who'd started losing teeth or who had lumpy udders. Cull sheep were doomed as mothers. Cull sheep would go for slaughter before we turned in the rams to breed the next year's crop of lambs. Cull trees were similarly troublesome and valueless. Cull trees were ones you couldn't give to a sawmill.

One question that I thought would surely be addressed in this comprehensive document was what our trees were worth—or what they might be worth if the timber market had been relatively strong. That didn't come out at all in the report, though. Probably a forester could make an educated guess, based on all the other data; for example, I now knew that forty-six percent of the basal area of dominant and co-dominant trees in Zone #1 consisted of sugar maples, and that the overall basal area per acre in that zone was ninety-five square feet. Also that most of the trees in Zone #1 were either midsize or large in size, and mainly AGS, that is, Acceptable Growing Stock. That sounded like good news; everybody knows that maple flooring is expensive. So are maple cabinets and maple furniture. Surely there was money there. But how much, I didn't know—and I kind of wanted to, if only to feel that the maples represented a potential source of wealth. What I did learn was that we had an awful lot of them—in Zone #1, anyway. But

I already knew that. Zone #1 had been a sugar bush before we bought the farm. Its trees had long been tapped for making maple syrup.

Most of the management directives in our plan, though, had nothing to do with anticipated timber harvests. Not in the ten-year time frame of the document. Rather, the practices that we were committed to revolved around maintaining habitat for bats. It must have been the strangest Management Plan that our county forester had ever been presented with, coming as it did on the heels of new guidelines for the state's program of tax relief. But sometime in mid-October, word arrived that the plan had been accepted. So now most of our open fields and forested land here—all but twenty acres that we chose to hold out— were enrolled in the state's Use Value Appraisal program, and I had high hopes that this would slash our yearly tax bill. That had been the starting point for this entire exercise: lowering our annual property taxes. But we wouldn't know for sure how much they had been lowered till the following August, when the next year's bill arrived. In the meantime, visions of phenomenal savings danced in my head. I was a happy camper; this had been a job well done.

About that time of year—midautumn—I started working up three cords of firewood to see our house through the coming winter. That would be as much as we had ever burned before, back in the days when we were using wood exclusively to supplement our passive solar heat. But I was determined to cut way back on fuel oil. I kept Brendan's tip about the buckthorn in mind; when I took my chainsaw to the woods and tried to cut one, though, I found the stuff discouragingly difficult to work with. The twisted, multiply branched and misshapen stems of the tree-sized specimens made the wood impossible to cut into straight sticks. And if they weren't cut straight, they wouldn't stack well. Then there were the sharp, spiny thorns that emerged from the bark at random intervals—like so many nails poking out to prick my skin. Working up this species into firewood would take some patience—and a pair of heavy gloves. And far too much of even a large buckthorn was comprised of twigs and branches too puny to deal with; consequently, there would be enormous piles of waste. After taking down a buckthorn or two and struggling to convert their nasty stems into firewood, I gave up and went to drop a proud white ash.

Each tree species has a characteristic shape, and the ease with which a person can fell a tree and cut it up has a lot to do with the dynamics of that shape. If you want to move a lot of firewood quickly, white ash is the way to go. A twelve-inch-diameter specimen at breast height is apt to soar for thirty feet before it even hangs a branch. There's very little top or crown to have to deal with, relative to the amount of wood that's in the trunk. Ash is dense enough to burn well, but a chainsaw zips through it like a stick of butter. And unlike some other hardwoods, ash has a straight grain that makes it split easily. I had learned these properties of ash trees forty years ago, as an eager back-to-the-lander with a hungry woodstove; now, as an aging refugee from academic life, I was about to refresh my memory. The ash tree that I chose was standing right along the forest's edge, and it wasn't hard to drop it into the adjacent field. Trees on the edge of a forest often lean out to gather extra sunlight; this makes them easy to bring down without risk of getting tangled up in other trees.

In my late twenties, I found that I could cut and haul and split a cord of white ash—maybe even stack it, too—in a single day. Not anymore, though. Now my back was aching after just a few hours of work. Still, half a cord per day was an attainable goal for cutting white ash. It saved a lot of time, compared with oak or sugar maple—let alone the buckthorn. Unfortunately, I was now acquiring the values—and the conscience, too—of a responsible forest manager. Was it wise to specialize in burning white ash? The tree had several high-end commercial uses: baseball bats, canoe paddles, tool handles. Hockey sticks. White ash was also used for furniture and flooring. And here I was creating a selection pressure against one of the highest-value species in the forest, simply because ash made a quick and easy firewood. I had just taken down a tree that a wiser person would have left growing till it reached maturity, then taken to a sawmill.

That train of thought made me turn to the deadfalls that were littering the forest floor—almost everywhere, it seemed. Many had been lying there since The Great Ice Storm; consequently much of the wood had gotten soft and punky. Sometimes not, though—or not beyond a mere patina of decay. Again, it had a lot to do with each tree's species.

Beeches and birches were definitely goners, but the stems of red and white oak were basically intact. Since the woods needed cleaning up just to move around in, I began cutting up some of the sound wood lying at my feet. It wasn't quite as sweet to deal with as white ash—*green* white ash, to boot—but I found I still could be productive with these deadfalls. And it would get some of the debris out of the woods.

The drawback was that I could work efficiently only within about fifteen yards of the forest's edge. I could park my pickup in the field close by, drop the bed's tailgate and toss in wood until the springs began to sag. Most of the actual labor was the carrying. But once the source of wood became too distant from the truck, the effort of being a packhorse wore me down. Forty years ago there had been trails through this stretch of woods, some of them wide enough to admit a truck or tractor. But they'd all been closed in now with new forest growth. So apart from just a narrow strip along the woodlot's edge, the trees weren't easily accessible for harvesting. Not with what I had for equipment, at any rate.

That was a depressing revelation, as we entered fall. But I'd cut enough fuel to see us through the coming winter, and I had a different sort of woodworking job in mind. I wanted to build a reasonably large and distinctive house addition—despite the fact that the house we already had was plenty big, for just two people. Why, then, add on to it? In a way, as a retirement present to myself. I was done with trying to be smart and academic; I wanted to see what I could do with my hands again. It was to be a freestanding house addition, joined to our main quarters by a covered wooden deck. Twenty-four by thirty-six feet seemed a nice size for what I called "The Annex," most of it committed to a large multipurpose space. We could put up guests there, if we liked—maybe rent it out to visitors from time to time. Cheryl could use it to host meetings and conferences on child-welfare issues. In the dead of winter, I could use it as my office. True, a lot of empty nesters like ourselves were downsizing—and I understood their reasons. But few things in life had given me the pleasure that construction projects did, and I knew I wouldn't have the strength to engage in heavy carpentry forever. Also I'd been testing certain radical ideas—such as rejecting conventional foundations—in some of the smaller structures that I'd put up on the farm. I wanted to try these concepts out on a larger scale.

There was something else, too. My father had passed away six months before—died at the age of ninety-three, so it was not a shock. But ever since, I had been dealing with some issues. When I considered my typical reaction to anyone's assertion of authority over me—a response that could probably be summarized as *Fuck you*—I was pretty sure I knew who had set the stage for that. My dad's assertion of authority had been extreme; if you disappointed him, he turned into a tyrant. He set many goals for me, then moved the goalposts ever further away. When I came to feel there was no way to satisfy his outsized expectations, I had rebelled against them. I rebelled against him, too. The fact that I'd "succeeded" after going my own way in life couldn't help but irritate him, even years later. Sometimes he was able to keep his rage in check, but at other times it would spill out in annoying ways. At times, his relation to me seemed like a person who keeps picking at a scab.

My father had been an electrical engineer—he had designed the thin sheet-metal laminations that go into transformers—but he also had a full set of construction skills. So did *his* father, who had died when my father was just a boy. So does my son, Ethan. And so, I'm proud to say, do I. It's an amazing fact that none of these fathers ever taught their sons how to build things—not in an overt way, at any rate. Carpentry, masonry, plumbing, and wiring—there were no formal lessons, no hands-on apprenticeships. As a child, I had been too busy with my own concerns to work beside my father on one project or another. Thirty years later when I had my own son to raise, it had been the same way. But when the time was right, in each person's case a hidden aptitude emerged. Kind of like a bat finding the way back to its cave.

Once I had begun to put up buildings on the farm, my dad loved coming out to spend time helping me. Father and son bare-chested, working side by side with our hammers and our saws. Some of my best times with him had been spent that way, after I'd grown reasonably competent with tools. (Criticizing other peoples' lack of competence was a big thing with him.) Now, in the wake of his death, a weight felt lifted from me—and I felt allowed to try to understand my father's psyche without fear of pain or punishment. Trying to honor the goodness that I knew was in him, while trying to figure out why he had to be so impossibly difficult. So angry and judgmental. And such a frightened and vulnerable man, if you ever got to duck behind his Oz-like curtain. I had lots of complicated memories of my father, but I'd kept them safely bottled. Now it felt like time to pull the corks and try to analyze them. And since my best times with him had been spent on building projects, I wanted to make my psychological investigation with a hammer in my hand.

To put up a building in Vermont with no foundation, you need a site that offers solid ledge beneath the ground. Thirty to forty inches down is an ideal depth, and I had that situation just a few feet from our house. You dig down to find that bedrock with a clamshell hole digger, then you plant a post in every hole and backfill it. All the building's framing members fasten to that set of posts—or fasten to another stick of lumber that is fastened to them. Building this way gets you out of the ground fast, and it saves a lot of money. True, at some point the posts will start decaying and create a problem that will have to be addressed. But I can think of several ways to address it. And by using pressure-treated posts on a well-drained site, serious decay shouldn't set in for a long time. Forty, fifty years or more. By which time someone else will have to deal with the problem.

Framing up a building is exhausting and obsessive work; once you get into a rhythm, days and weeks fly by. My goal was to get the structure closed in by the start of winter, then spend the cold months doing the interiors. Incredibly, I was building without any written plans; all of the construction details were in my head. Materials were rock-bottom cheap, given the housing meltdown. Sometimes when I'd show up at the lumber yard, the workers cheered. Sure enough, before first snow had fallen I was hanging drywall in a tight and heated space; by Easter of 2010 we had a small party to inaugurate The Annex, which had come in on time and well under budget. It seemed a fine achievement for a man of sixty-two, and I felt light-years from my gig in academia.

Not that I had figured out my father during those six months, but at least I'd been communing with his workaholic spirit. He had been compulsive, too, about getting projects done. Without a project to invest in and obsess about, my father quickly grew depressed. I was that way, too, I now couldn't help but realize. Luckily, as soon as the addition was completed there was farm work to turn to. The sheep were ready to start dropping that year's crop of lambs; then we would be into pasture management and haymaking. Oddly enough, although we now had a state-approved Forest Management Plan it didn't seem like I had much to do in the woods. I had hardly set foot in them since cutting firewood the previous fall. "Maintain all shagbark hickories"—that had been the main directive of the document that Vermont Family Forests had drawn up in our behalf. Maintaining them seemed to mean doing nothing, as far as I could tell.

All that was about to change, though. We were about to be invited to another party put on by the government, and to get another windfall from our fellow taxpayers. WHIP was the bright and clever acronym *du jour*, derived from something called the Wildlife Habitat Incentive Program. Certainly this sounded like a noble and worthy cause; why not incentivize the habitat for wildlife? But WHIP had other meanings that were not at first apparent.

The WHIP

Early in June of 2010, Brendan Weiner got in touch to say there might be federal money we could tap to help begin daylighting the shagbark hickories—opening up the forest around likely roost trees in ways that would help them to attract a few bats. Or attract more of them, if bats were there already. The money would come from the Natural Resources Conservation Service (NRCS), a division of the U.S. Department of Agriculture. They had many programs handing out cash to qualifying landowners. WHIP was a relatively minor initiative, but one that was being well-promoted in Vermont. If you were a landowner seeking to enhance certain kinds of wildlife habitat, WHIP would offer technical assistance and then pay you money on completion of their specified practices.

It seemed reminiscent of my experience "co-operating" with the U.S. Soil Conservation Service thirty years earlier, so that set off alarm bells. Then I discovered that this Natural Resources Conservation Service outfit *was* the U. S. Soil Conservation Service—the exact same agency as I had dealt with to design our farm's pond and build diversion ditches. The name change had occurred back in the 1990s, under President Bill Clinton; "Natural Resources" implied a broader mandate than mere soil concerns, and this expansion of the mission seemed justified in view of how the world had changed. Clean air, pure water, biodiversity—even climate change was on the agency's revamped agenda. Though only a small division of the nation's vast Department of Agriculture, the NRCS employs about twelve thousand people. In Vermont alone—a tiny state, after all, in terms of size and population—there are roughly one hundred agents organized into a central office and eleven field offices. Nearly every county has one. Brendan referred me to a conservation specialist—one George Tucker—at our own county's field office, located in the same building as the county forester, the University Extension Service and other agencies of government whose job it is to help farmers. Or at least to "help" them. If I checked this WHIP program out and it looked attractive, time was of the essence; Brendan said there was an application deadline soon. Once we got that filed and open for review, it could take a year or more to get awarded funding.

I took Cheryl along to meet with George Tucker, since I expected there'd be paperwork to do. She's quite calm and efficient with a sheaf of forms, while I tend to go berserk and start acting out. George was an affable and interested bureaucrat; he pulled out our Forest Plan and riffled through its pages. The basic application for entering WHIP was only three pages long, although it came attached to a fourteen-page appendix. We could read that later, though, since it wouldn't actually be pertinent until we had been offered a contract. There were certain declarations that we were required to make—for example, that our land was privately owned and that we could prove we were the people who owned it. That made sense to get nailed down, I figured. Also, we had to swear that our adjusted nonfarm gross income did not exceed one million dollars per year. No problem there. If it had, though, that circumstance wouldn't necessarily have disqualified us from feeding at the public trough; there would just be several other forms to fill out. Then there were some questions about whether we might qualify as Beginning Farmers, Limited Resource Farmers, or Socially Disadvantaged Farmers. We felt we were none of these. If we'd felt otherwise, we'd have to meet some "self-certification" requirements—and it seemed like doing so would likely grease the deal. Why else would they have asked? Not that the bats would have cared, it occurred to me.

All in all, it was a fairly painless interaction with an agency of government. We were not committed to doing a thing, yet, and the agency's commitment was only to deciding if they felt inclined to help. We signed the application, shook hands with Tucker, and were on our way with a "don't call us, we'll call you" understanding. Back at the farm, we took a hike up the back cliff and stared down at our domain. Fields, forests, pond, and wetlands intersecting with each other. Almost holding hands, it seemed. Maybe things weren't quite perfect down there, but from up on high the land looked lovely. Even after forty years, its beauty took our breath away. Something fateful was about to happen, we both realized. We were about to get entangled with the government.

A few days later, the phone rang and I spoke with a resource conservationist from the statewide Natural Resources Conservation Service office, in Colchester—an hour's drive north of here. His name was Toby Alexander, and he sounded eager to make a site visit in order to inspect our woods. Our request for WHIP funding had gotten fast-tracked; when I told Brendan, he said that such a quick response was virtually unheard of. Evidently, the bats' worsening plight had made them a top priority—and how many other landowners could there be whose Forest Management Plans were aimed specifically at helping bats? Not too many, I figured. So there was a fit—a harmonic convergence, even—between our stated purposes and those of the WHIP people. Sure, I said to Toby Alexander. Come anytime you like and have a look around.

The resource conservationist arrived on a sweltering afternoon a few days later. Hazy, hot, humid; it felt more like the South Jersey of my teenage years than the Green Mountain state that I had managed to escape to. That was global warming, though. Or climate change, or whatever. Toby was dressed for the heat—T-shirt, baseball cap—but I didn't envy him a trek through the woods on such an oppressive day. Then he had to strap on a GPS device, which hardly looked comfortable. The man from the government exuded a professional competence, along with what struck me as a sense of formal distance. He wasn't trying to make friends or shoot the breeze with me. There seemed little point in trying to chat him up. He set off exploring, and a couple hours later he returned soaked in perspiration. It was like he'd stepped out of a sauna, or a Turkish bath. "Well?" I asked hopefully.

"You have a lot of nice roost trees, back there. But you also have invasives."

"Ah," I said. "You found the buckthorn. I've been meaning to get on that."

"I found garlic mustard, too."

I was clueless. "Come again?"

"White flower. Heart-shaped leaf. Grows in a rosette."

"Wait. Is this a tree we're talking about?"

"More like a wildflower. Not one that we like to see, though."

Smartass, I thought to myself. A hotshit botanist. I was pretty sure that I knew all the major tree species growing in our forest—knew them by their bark, their leaves, their seed pods and their branching patterns. Was I supposed to know the wildflowers, too? Cheryl did know most of them, and she'd even taught their common names to both of our kids. But ground cover in the woods had never earned my interest. "So is that a problem?" I asked, a tad defensively.

"It is, as far as we're concerned. If you'd like to move ahead with funding from WHIP, it needs taking care of."

This did not make sense to me, but I wasn't ready to get into a fight. Yet. Toby said that from his agency's perspective, dealing with invasives had to go hand in hand with improving wildlife habitat. Sort of a fair warning, when I look back on things. Then he said he'd like to take a walk around the woods with the person who had drawn up our Forest Management Plan; he wanted to know exactly what Brendan had in mind when he had prescribed daylighting the shagbark hickories. Alexander hadn't dealt with Weiner before, so I gave him Brendan's contact information at Vermont Family Forests. Then he said that even better would be to get Brendan here along with one of the state's people. Maybe Jane Lazorchak from Fish and Wildlife, since she had been in on this project from the start. Fine, I told him. LIP and WHIP—it sounded like a marriage. But I said I'd like to walk around with them, too, learning what I could about this contract we were heading into. Toby said if everyone was on the same page, things could move fairly quickly. His office had some uncommitted funds for the current year, and bats were a matter of urgent concern. Then he got into his truck, cranked the AC, and headed back to Colchester.

On the whole, I felt pretty good about the situation—almost as good as I had felt when Scott Darling asked if he could trap some bats here four years earlier. White-nose syndrome was now being called a natural catastrophe, likely to profoundly alter the ecosystem if nothing was done about it. If we lost the bats, then the insect populations that they fed upon could burgeon. That could have dire effects on farmers' crops—plus, we'd all be spending lots more time swatting mosquitoes. Trees could be denuded by leaf-eating bugs. We had a serious problem on our hands, but I was strategically positioned to help solve it. Or at least participate in finding a solution. And I'd been discovering that turning conversations to "the bat project" here was bound to win attention at any social gathering. Win fond looks from admiring women, too. If I'd been a single guy, I could have gotten dates with it. Heck, if I'd been looking for love I would have had my hands full. In up to my neck. What could be more sexy than a guy who owned a forest and was using it to save the bats? I was like an eco-hero. I could come on like an environmental activist, and all I'd really done so far was sign a couple forms.

At the same time, I was starting to build a philosophic case against freaking out over so-called "invasives." Or "exotics." Or "exotic invasives," to fire both barrels. Why make them such a big deal, after all? Every living thing on earth, if you took a long view, must have at one time been invasive in its ecosystem. It was a matter of survival of the fittest. Now that I knew what glossy buckthorn looked like, I couldn't help but see it everywhere I went. All along the roadsides of our county, for example. And at all the nearby state parks, which were managed by the same branch of government that hired county foresters. There was buckthorn doing well at all the posh ski resorts, and along the hiking trails scattered through the Green Mountain National Forest. At some point, we'd have to stop calling it invasive and admit that it belonged here. That the plant had earned its place. The buckthorn seemed to think it did—that much was for sure.

And speaking of invasives, what about the Anglo-Europeans who had settled here beginning in the 1600s? They did not "belong" here; they had come from somewhere else. Hadn't they spread rapidly, causing widespread damage to existing ecosystems? Didn't they squeeze out a host of former native species? Weren't they way, way too aggressive for the competition? Four hundred years is not a long time, in the world of nature; the Anglo-European "they" was actually we ourselves. But if we were poster children for the concept of invasives, who were we to call the shots when it came to glossy buckthorn—or this garlic mustard plant? Wasn't that the pot trying to call the kettle black?

These thoughts didn't play too well a week later, though, when a whole team of experts came to walk around the farm. There was Toby Alexander from the state headquarters of the Natural Resources Conservation Service, along with George Tucker from the local field office; there was Jane Lazorchak from Vermont Fish and Wildlife, and Brendan Weiner acting as my forestry consultant. And my ally, in case we got into an argument. Also there was Yuki Yoshida, a Middlebury student who had recently graduated in environmental studies—she would be on my side, too, if a fight broke out. And then there was me, the nominal reason for this pow-wow that was costing the taxpayers a lot of money. Pickup trucks rolled up the driveway, one by one; people put their hiking boots on and donned their field gear. Then they all shook hands and began a friendly pissing match, probing to see which person knew the most about bats. Or who knew the least about them.

Almost immediately, Brendan found himself under a fairly harsh interrogation by the men from WHIP. I watched him keep his cool; he stood by the Management Plan he had drawn up for us, and by the thought and research that had gone into it. He had certain *bona fides* that he called attention to. Vermont Family Forests was one of only two consulting forestry outfits in the county who were qualified to draw up such a management plan, based as it was on wildlife considerations. If someone wants to make the welfare of forest creatures—bats, in this case—their woodland's chief priority, this is the kind of Forest Management Plan you get.

At some point, I suggested that we head into the woods. Once there, we started picking out the bigger hickories and talking hypothetically about which trees around them people thought should be removed. Everybody seemed to have a different idea. Just a few months before, Vermont Fish and Wildlife had published formal guidelines—written by Scott Darling, I later found out—to help people make exactly these decisions; since the guidelines had been appended to our Forest Plan, everyone had had a chance to read and consider them. When it came to actual cases in the field, though, the document was manifestly open to interpretation.

One principle was clear: a shagbark standing by itself in an open field had no value as a roost tree, since its exposed position put the bats at too much risk when they ventured out. Bats knew better than to make themselves sitting ducks for owls and other predators. So "opening" the forest to that extent would be mistaken. Bats need a degree of cover—ample canopy within twenty feet, say, of a likely shagbark hickory. But the bats also like their roost trees to get lots of sunshine, helping them stay warm. The twin goals of sun and shade were frankly incompatible, so there were some calls to make. Tough, complicated choices. We tramped together from one shagbark to the next, and each of four experts would propose a different nearby tree they felt would have to go. There was an art to making these decisions, and the experts who had gathered here did not always agree.

Then someone knelt down and plucked a certain plant out of the ground by its S-shaped root. It was Jane Lazorchak, I think, who couldn't stay for long because she was a new mother. She had a baby to get home to and feed. Actually, it wasn't just one plant but a family whorl, growing at our feet amid a carpet of its buddies. White flowers, heart-shaped leaves. "You know what this is?" she asked me.

"Let me take a wild guess. Garlic mustard?"

"Yes. And you have a fair amount of it." The situation here was not a flat-out infestation, but she told me that would happen quickly if I took no action. Jane crushed a leaf or two and held them underneath my nose; I caught a faint whiff of garlic. And then mustard, too. The plant was a biennial, she told me, with two distinct iterations in the course of its career. The low white flowers huddling near the forest floor belonged to year one, an innocuous stalking horse for what was yet to come. They looked passably attractive as a ground cover, and it would be easy to mistake them for a native plant. A harmless wildflower, like trillium or trout lily. Something that belonged here. But in year two, each plant would suddenly go crazy and produce a three-foot stalk. Jane now searched to find one of these, and pulled it out. The upper reaches of the second-year plant went into a kind of seizure, throwing out seed pods or *siliques* in all directions. Spaced along each tan and papery pod was a row of tiny black seeds, waiting to be cast adrift and rain down on the nearby soil. Thousands of seeds, I'd have to say, on just this single plant.

George Tucker took the seed-bearing plant and hung it from a tree branch, so its roots would desiccate; if you just tossed it on the ground and walked away, he said, it might succeed in getting reestablished. Toby Alexander frowned. Hanging garlic mustard out to dry was a mistake, he said. The seeds were already viable for next year's crop, and sooner or later they would fall to the ground. In fact, he said the seeds could stay viable for *five years*, waiting for their chance to make firm contact with the soil. When and if they did, watch out.

"Why is this a problem, though?" I asked with what I hoped would pass for common sense. "How is garlic mustard going to interfere with bats?"

From the way they looked at me, clearly I was missing something. "Once you start to open up the forest," explained Toby, "there'll be more light penetrating to the ground. Sunlight, warmth, plus the disturbance of some logging—that's what the invasives are waiting for. Give them their conditions, and they're likely to take over."

"But they won't be taking over hickories," I said, digging in. "Bats don't roost near the ground, anyway. They like being high up."

"Yes, they do. That's not the point, though." The experts weren't with me; even Brendan Weiner was unwilling to take my side. *First do no harm* seemed to be their starting principle, and taking any action that might favor the invasives was perceived as doing harm to the forest's ecosystem. Garlic mustard, once it got aggressively established, could drive out the other flora living on the forest floor. That would in turn affect the animals who ate those plants. That would in turn affect something else, and so forth. Everything was interconnected, did I see? Before making changes to the forest's present character by daylighting the hickories, I would need to get on top of these invasives. Even run them out of town, if it were possible. Ninety percent control was the standard goal. Otherwise, we'd only make a bad situation worse.

By now, I was feeling like my woods had caught some dread disease. Leprosy, perhaps. Or herpes. And the infection had erupted on my watch—except that I had not been watching. I had been preoccupied. Whenever someone sets out to try and make me feel bad, they generally succeed; this is a product of my childhood upbringing. I can be made to feel guilty about things that have nothing to do with me. Isn't that a precious gift? In Sunday school, we used to sing a song called "Oh, Be Careful" and I took its words to heart:

Oh, be careful, little hands, what you do.
Oh, be careful, little hands, what you do.
For the Father up above is looking down with love
So be careful, little hands, what you do.

This was followed up with verses about what your little eyes might see, what your little ears might hear, and what your little mouth might say. Possibly what your little nose might smell, given the opportunity. The gist of it was that all these organs of sensation had minds of their own and might run amok on you—and that if you let them do that, God would stop loving you. That's a heavy load to put on five-year-old kids, but probably most of them were smart enough to disbelieve it. I didn't disbelieve it, though. I was credulous. I used to lose sleep worrying about what my little hands might do—or might *not* do, if they should have done one thing or another. And here my little hands—grown large, but much the same—had not been pulling up the buckthorn and the garlic mustard.

"Look," I blurted out. "I'm really sorry about this. Ever since the Ice Storm, I've stayed out of the woods."

"What ice storm?" asked Yuki, who is from Japan.

"It was like ten years ago."

"Twelve," said George Tucker.

"Don't beat up on yourself," said Jane Lazorchak. "We see situations like this all the time. It takes a lot of work to hold back invasives. Just because they're here doesn't mean it's your fault."

At that point, I had a thrilling revelation. These people were not my parents, after all; they were not my schoolteachers or scoutmasters or bosses. Hell, they were all a lot younger than me. I had twenty years or more on every single one of them. Why was I apologizing? I used my little hands to pull up a rosette or two. "Whose fault is it, then?"

"You can't stop the birds from eating seeds, or the wind from blowing. That's how these things get around."

"That, and all the farms that have gone under," someone else put it. "Once the open fields start reverting to brush, everyone's in trouble."

"But you'll have to get on this before you start cutting trees," Toby Alexander said. "We'll need to see that you're on top of these invasives."

Whose forest is it? I thought darkly. But I got his gist. If I declined to sign a contract with the government, I could do whatever I wanted in my woods—so long as it was generally consistent with our Forest Plan. But if I was interested in getting paid by WHIP to do good things for bats, I'd have to do it their way. "How do I control them?" I asked neutrally. Noncommittal.

"Manual removal works best, for the garlic mustard."

Manual removal—I had to process that. "Pull them out by hand?"

"Nothing else is as effective. Be sure to get them by the roots, so they won't grow back. Then put the plants into plastic bags and seal them tight. Don't try to compost garlic mustard, or the seeds will sprout. Don't ever take them to a landfill, either."

"Why not?"

"Because the seeds will sprout there. And that would make things worse."

"So I just hang onto them?"

"Put them into black plastic bags, and leave them in the sun. When enough heat builds up in there, the plants will die."

"Even the seeds?"

He shrugged. "The seeds can last for five years." Then he reminded me the plant was a biennial. Pulling out a plant in its first year would prohibit it from entering its seed-production phase, but it would take two years of control to get on top of things. As for the glossy buckthorn, very young specimens could be pulled out by hand—roots and all, that is—especially when the ground was wet. Anything bigger, though, would have to be cut off with pruning loppers. Or, for tree-sized specimens, a chainsaw. "Cut it off around eight inches from the ground," said Toby. "That way if it sprouts on you, you'll have enough stump left to cut it off again."

"Wait. It's going to sprout from a dead stump?"

"The stump's not dead. Not until you paint it with a glyphosate product."

"Glyphosate?"

"You know—like Roundup. Though we can't recommend any particular product."

Now I felt a wave of resentment—now they'd punched my buttons. I am not a dyed-in-the-wool organic farmer; if one of my sheep gets sick, I want the right to medicate her. But I'd never messed around with broad-spectrum herbicides, and I didn't even want to have them on the farm. Why was the government pushing glyphosate on me? Why was the Natural Resources Conservation Service pimping for Monsanto? "Roundup," I said. "Actually, I've never used that."

"Well, you're going to have to if you want to kill the buckthorn. We like to see it used at twenty-two percent. And it only works in autumn, when the plant's nutrients are moving down into the roots."

"Don't I need a license to apply systemic herbicides?"

"*You* don't, since you own the land you're going to use them on. If you were to hire someone else, though—yes, they would."

I turned to Brendan. "Are you a licensed applicator?"

"No."

I said I'd need some time to think this over; that was okay with the government's men, although I think they probably saw me as a wuss. Glyphosate wasn't like arsenic, after all. It was not plutonium. It just had a way of killing any growing plant it came in contact with—killed it dead by interfering with a basic enzyme process. Since the stuff was nonselective, you had to be careful with it. Very, very careful. But if I was going to get a leg up on the buckthorn problem, painting glyphosate on their stumps was the way to go—short of investing in a backhoe to dig them out. And if I refused to do battle with the buckthorn, the government would refuse to spend its money on me. There would be no contract to help bats if I would not play ball.

As our exploration of the woods dragged on, the experts found a couple of other unwanted species—though nothing on the order of the buckthorn and the garlic mustard. Barberry was one of them: a yellow-flowered, spiny-branched shrub held to be exotic. The Japanese variety—of which they found some specimens—was held to be particularly troublesome in "open" forests. Daylighting the hickories would make our forest open, too. Toby found some barberries near Cheryl's *poustinia*, and he duly tagged a bush with flagging tape for future reference. And then he found a bush of honeysuckle, too—another flowering shrub that he said could be big trouble. But not all honeysuckles in Vermont were exotic, and it took some skill to tell the ones that were from those that weren't. One way was to sever the bush's central branch and see if its inside was hollow or not; hollow meant exotic, and solid meant a native plant. But that seemed to me like finding out who was a witch by tossing suspected women in the nearest creek—to see if they would drown or not. Right, I said. Honeysuckle. We flagged one or two of those, so I could find them later on and learn to identify them.

Our field trip had now begun to give me a headache, but we kept at it for at least another hour. We wandered into precincts of the forest where I'd never been—not in thirty-eight years of tax-paying ownership. We saw places that were dense and tangled, places where the trees were getting choked by climbing vines. We were on the hunt for shagbarks, but it wasn't like they were distributed evenly; they tended to show up in clumps and clusters. Family groups. There were not invasives growing everywhere, either—and thank God for that. Much of the forest was essentially free of them, especially where the canopy was thick enough to thoroughly shade the ground. The strategy that I could hear the experts working out was to choose shagbarks growing where invasives seemed controllable. That was where it made the best sense to start our project. When they found such conjunctions—well-grown shagbarks, and controllable invasives—someone would record that location with the GPS.

Finally, after we'd all started dragging from picking our way over downed limbs and fallen trees, we hiked down the snowmobile trail to find the spot where Scott Darling had trapped his bats in 2006. He had recorded the relevant coordinates, and they were available to wildlife biologists. So we all stood on the spot where this odyssey had actually begun, and I watched the experts scratch their heads. There wasn't a shagbark hickory in sight. If that really was the bats' favorite roost tree, they could have done better elsewhere on the farm. Then I pointed out that Scott Darling had surmised that bats might be using the trail as a flyway. It headed more or less straight to the farm pond, where there would have been a plenitude of bugs to feed on. Maybe that had drawn the bats—the quality of foraging.

That turned the conversation to a related issue: trying to provide better habitat for bats to *feed* in, as opposed to finding roosts. We had been so focused on their need for shelter, it was like we'd overlooked their parallel need for food. Brendan and the men from WHIP tossed around whether we should try to do that, too. There were places in the woods adjacent to the pond where some aggressive thinning might make it easier for bats to fill their stomachs. Turn an objectionably dense, crowded patch of forest into what could serve as an all-night diner. Or a food court, someone joked. Here's your fly-through Bugger King, here's your McMosquito's. We were getting slaphappy; it was time to call it quits.

Back at the house, people stripped off their field gear and got into their office clothes. Toby Alexander told Brendan that he'd be in touch; after the government men had departed, Brendan asked to come inside and talk with me privately. We opened beers together. "You okay with this?" he asked.

I wasn't sure, but there was one piece of information missing in my calculus. "How much money are we talking about here?"

"They have a list of rates per acre to do certain things. Maybe like a hundred bucks an acre to control invasives, and a hundred more to cut particular trees. I'm just guessing, though."

"They don't pay by the hour?"

"No."

"Why not?"

Brendan shrugged. "They have their policies."

"But some of those acres would take lots more time than others."

He didn't disagree. "That's the way it is," he said. "At least they're willing to pay a little something."

"How bad do you think things are, with the invasives?"

"Buckthorn is the worst of it—and I knew they'd jump on that. I didn't think they'd get intense about the garlic mustard. On the other hand, you're lucky. They could have called you out for having Norway maples."

"Do I?"

"Yes, you do. A lot of them."

"So why didn't they?"

"They might not have recognized them. Norway maples are hard to tell from sugar maples—which of course do belong here. What you have to do is take a leaf and squeeze the petiole. If it's a Norway maple, you'll see a drop or two of milky sap."

"And if it's a sugar maple?"

"Then the sap will come out clear. Otherwise, the two trees are really hard to tell apart. But these Norway maples—"

"Came from Norway."

"Right. And if they'd told you to get rid of those, you'd have your hands full."

The way we left things, Brendan would wait to see what the men from WHIP proposed. If they did offer a contract to do forest work, it would have the rates of compensation clearly specified. If I didn't want to do the work all by myself, I could use the government's money to pay someone else. Maybe hire Brendan, in fact, to cut the buckthorn down—and then I could work behind him painting stumps with glyphosate. Maybe hire local kids to go after the garlic mustard. That was the sort of work a teenager could do.

This idea called up memories; I used to do work like that. As a teenager in South Jersey, there had been a plague of Japanese beetles—copper-colored insects that had come here from Japan. All over our prim, well-manicured subdivision, people were hiring kids to strip the beetles off their trees and drown them in gasoline. One cent per beetle, and it added up fast. Kids running around suburban backyards with Mason jars filled with gasoline and half-dead beetles had to be a nightmare, but there was no Environmental Protection Agency in those days to police it. This was in the late 1950s, and the early sixties. Everyone was happily innocent about pollution. Gasoline had been our glyphosate, so to speak. And nobody was trying to make sure we used it safely. For that matter, housewives were killing ants with spray bombs laced with DDT. Right in their own kitchens, where they cooked their families' food.

I thanked Brendan for hooking me up with this program called WHIP, but I told him I had reservations. Cheryl was out of town—in DC, attending a government conference on one child-welfare program or another—and I wanted to bring her up to speed. I knew that she wouldn't be too keen on using glyphosate. Brendan left me brooding—and the more I thought back over the day's events, the more I felt upset. And wary, too. What had started out as a free pass to tax relief seemed to have taken a wrong turn somewhere. I was on the verge of taking orders from the government. Do this, do that, shape up, fly right—I could hear my father's voice coming through loud and clear. And as I had learned with him, I sensed that these agents of government authority might be very hard to please. If I bent over and did everything they asked me to—pulled the garlic mustard out, painted all the buckthorn stumps with broad-spectrum herbicide, grubbed out the Japanese barberry and honeysuckle—there was no guarantee that Toby Alexander and George Tucker would be satisfied. They might find some other noxious plant to rub my nose in, something that none of them had noticed before. Like those Norway maples, say. They might pull a leaf off a tree and squeeze the petiole, and totally freak out. And if I went along with their arcane directives, it would be two years before I actually did something that would benefit the bats. Two years of waging a war against exotic plants—by then, who could say there would be any bats left to save?

I did a web search to bone up on glyphosate. Yes, it had originally been Monsanto's product—Roundup—but this famous cash cow had long since gone off-patent. Now it was marketed by several giant companies under a host of names. Ranger, Killzall, Rodeo, Eraser—all were glyphosates. Lumping the various products together, glyphosate was the most widely used herbicide in the United States. Two hundred million pounds per year, in round numbers. On a toxicity scale from I to IV developed by the EPA, glyphosate scored a III—with category I being reserved for the worst offenders. So there was more poisonous stuff out there, for sure. Glyphosate was said to begin degradation shortly after application, sometimes in a matter of just a few days. Since it became tightly bound to soil particles, groundwater pollution was not a huge concern. One problem in forest use was that the product sometimes migrated from one plant's root system over to another's. Glyphosate's action was completely indiscriminate, killing any growing plant it managed to connect with.

Next day, I took a walk back into the woods and tried to find the stumps of buckthorns from the year before—the ones that I'd cut down as a firewood experiment. They weren't hard to find; each stump had sprouted a phenomenal wig of shoots. Tall, green, and healthy-looking. Some native hardwood trees are able to send up new sprouts from a cut stump, but nothing like this; the buckthorns were like zombie plants, essentially undead. Then I found some baby buckthorns growing nearby—the kind it would be easy to dismiss as a harmless weed—and yanked with all my strength to uproot them from the soil. A buckthorn plant no bigger than a pencil aboveground had a root system the size of my fist. The roots were black and dense and dazzlingly articulated, branching and rebranching like the trees themselves did. Curious, I tried to pull a modestly larger plant—this one about the size of a baton. When it wouldn't budge, I went home to get a shovel—just so I could see what sort of critter I was dealing with. That one's root ball was the size of my head. Thinking what the tree-sized specimens must have lurking underneath the ground was more than disheartening; as Brendan had told me when he'd first raised the issue, buckthorn was a plant whose survival strategies outstripped the competition's.

Cheryl came home the next day, and I laid things out. WHIP would like to pay us good money to cut trees, I said. To release the shagbark hickories from the neighbor trees that shaded them, or to daylight them—once the program's agents had agreed on which trees ought to go. But when they did go, the roost trees would attract more bats. First, though, we'd have to kill the buckthorn and the garlic mustard. (I didn't bring up the Japanese barberry or the possibly exotic type of honeysuckle.)

"Why?"

"Because they don't belong here."

"Why not?"

"They're invasives."

"So?"

"Look," I told her. "I don't understand it, either. This is what we get for doing business with the government." And then I said that getting rid of the buckthorn would mean painting their stumps with a product called Roundup.

"That's some kind of herbicide, right?"

I nodded. "Broad spectrum."

My wife rarely uses salty language, but she did now. "No fucking way," she said.

"Else the buckthorn comes back up. It uses the existing roots to sprout from the stumps."

"What does any of this have to do with helping bats?"

"I don't know," I told her. "But these WHIP guys are frothing at the mouth about invasives. If we don't get on them, they could get out of control."

"So?"

"They're bad for the woods. For the forest's ecosystem."

"Isn't Roundup bad for the ecosystem, too?"

I was reminded of a statement that an army spokesman put forth in Vietnam, which came to undermine the war there: 'To save the village, it was necessary to destroy it.' That was what the government seemed to be proposing—at least from Cheryl's point of view, and probably my own. Poisoning our lovely woods in order to save them. And from what? From plants that really wanted to grow there, but in ways that were too eager for the forest's good. I told Cheryl about the baby buckthorns I'd dug up—how their root systems were like nothing

I had ever seen. I told her how it could emerge in dense thickets. But she held her ground, and I could see her position. There was no reason why we had to go along with WHIP in order to help the bats. If the government's directives went against our values, so be it—the government could damn well screw itself.

That's where we left the conversation, for the time being. But a few days later, Brendan Weiner—who was, after all, an eco-minded forester—sent me a fact sheet on invasives that had been prepared by Vermont's chapter of The Nature Conservancy, an organization whose reputation for Green thought and practice was unimpeachable. Sure enough, Roundup and similar products were on their list of treatments for eradicating buckthorn. And for barberry infestations, too. By painting glyphosate on a freshly cut stump, its action would be focused on just that one plant. Unlike working with a foliar spray, the killing would not be uncontrolled and indiscriminate. They had a nonherbicidal treatment, too: you had to wrap the buckthorn stumps in sheets of heavy plastic, tied in place with baling twine. Then you had to take the plastic off periodically to check those stumps for sprouts—every few months for a couple years. Or longer. If you found that shoots were growing, you'd have to cut them back and reattach the plastic covers. That seemed like a lot of work, considering the number of stumps that I'd be dealing with. Hundreds of them, easily. Possibly thousands. And any stumps I missed would land me back in trouble. One glyphosate treatment ought to do the trick, though. And in The Nature Conservancy's calculus, the downside risks of using glyphosate for this purpose were more than offset by ensuring that the plants were dead.

So that got me thinking maybe Roundup was the way to go, all things considered. Meanwhile, on the computers and workstations of government officials, the bat project here continued to get fast-tracked. Aside from the problematic presence of invasives, Toby and George and Jane had liked a lot of what they'd seen. Parts of our forest had considerable bat potential. Plant a shagbark hickory seedling, and you're talking fifty years—at least—for it to grow into a viable roost tree. But we had scores of them that were good to go, in terms of size and height and elaborately shingled bark. Most of them were spread across a site that had a favorably south-facing aspect, nearby sources of water for the bats to drink and lots of nice edge effect where woods and fields intersected. And we had another asset almost no one else had: a Forest Plan that made the bats our number one priority.

About a week later, I found myself copied in on a long e-mail from Toby to Brendan laying out a plan. The idea was to demarcate three half-acre management zones—each of them roughly circular in shape—plus one long, narrow zone encompassing another acre and a half. The zones were superimposed on a satellite map enclosed as an attachment, based on GPS coordinates taken when we walked the woods. Toby's advice was to spend two years on "invasives control" in the designated areas, after which the forest canopy could be opened up near particular shagbark hickories. To tackle the invasives in a wider way than that, he felt, would probably lead to frustration and defeat. The garlic mustard could be pulled at nearly any point in the growing season, so long as the soil was loose enough to get the roots. But taking on the buckthorn and the other exotics had to wait until autumn—after Labor Day, at least—when painting glyphosate on their freshly cut stumps would likely assure a kill. By "freshly cut," Toby meant within five minutes. The longer you waited, the less the effect. Oh, and there was one more thing: when it came time to actually fell some trees to daylight shagbark hickories, we could only do it between November 1 and March 31. During Vermont's long winter, that is to say, when the roosting bats would have decamped to hibernacula. During the warm months, if any bats were in the woods the physical impacts of logging might disturb them—and if that happened, it could frighten them away. That was a new idea: people scaring bats.

As for the payments that WHIP was prepared to make, I was happily surprised. The standard rate for Year One of tackling invasives was two hundred thirty dollars per acre, and the rate for Year Two was one hundred thirty per acre. Then to daylight hickories so they'd better serve as roost trees, the going rate was two hundred dollars per acre—and I was welcome to cut up the trees taken down and use them as firewood. Two cords per acre was a reasonable estimate, and firewood was selling for three hundred bucks a cord. Eighteen hundred dollars' worth of firewood, then. Of course, that didn't factor in the time I'd have to spend to do it. But it was work that I'd been hankering to do. Forty years ago when we had purchased the farm, we'd paid a little under four hundred dollars per acre—though the value of the *forested* land was surely less than that. Maybe only two hundred dollars per wooded acre. Adding all the numbers up, the payoff for doing business with the government came to many times per acre what we'd bought those acres for. And we would still be the people who owned them. And they would be all cleaned up, ready to enjoy.

I showed Cheryl the going rates for forest work, if we entered into this contract with the government. She was suitably impressed. Then I showed her The Nature Conservancy's fact sheet on invasives—sent to me by our eco-minded forester—with its clear suggestion that glyphosate was a defensible way to get on top of buckthorn. That using it actually made environmental sense, if applied as recommended. "So," I said, "I'm starting to feel okay with this. How about you?"

"Do what you want," she told me. "I'm not going near that stuff."

I said that the buckthorn had to wait till autumn, anyway, but that it was not too soon to pull the garlic mustard. When I said that anyone could yank this weed out—even young children—Cheryl mentioned that our neighbor's niece and nephew would be arriving soon—from their home in Baltimore—to spend time with their aunt and uncle. They were good kids, and they seemed to get along together. They had helped us bring in hay, the summer before; afterwards, we had all gone swimming in the pond. So there was a source of child labor in the offing, if these kids were looking for some work to do. A project to keep them busy and make a little money. They were in their early teens, old enough to tackle the sort of task we had in mind without much supervision. If we could entice them into picking garlic mustard, they might get a sizeable chunk of the weeding done. What it would come down to was having an effective pitch—but it seemed our grown-up son, Ethan, might be up for that. He was a gifted teacher of young teenagers. He had a way with them. What's more, he knew Baltimore; he had spent a year there running work camps for students on behalf of Habitat for Humanity. So the pieces of a plan were falling into place.

Brendan wasn't fully satisfied, though. He told me he was holding out to have the feds add another acre to the deal. Again, he raised the issue that our habitat concerns should not be focused solely on providing bats with trees to roost in. Helping them to find some food should be on the agenda, too. There were the potential insect foraging areas that had been discussed when we'd walked the woods with agents from WHIP and LIP—places where the forest could be thinned to make an all-night diner. The areas that Brendan had in mind were near the farm pond, and not far from the snowmobile trail where Scott Darling had trapped his Indiana bats. What about it? Brendan asked the government in several e-mails, copying me in. Habitat means shelter, but it also relies on the availability of food. Why work to give the bats one, but not the other? Eventually he got Toby to see things his way, thereby juicing the deal on my behalf. Four acres, then—with the extra acre to become a foraging zone. All things considered, the value of doing business with the government would come to just over twenty-two hundred dollars. Plus eight or nine cords of firewood to heat our house—enough to last for three winters, if they weren't too cold and snowy. Firewood that the nation's taxpayers would, in effect, be paying me to cut for myself. A freebie. That in itself was a powerful incentive.

Looking back, I wish I'd gotten started rooting out invasives right away—even before signing my name on the dotted line. Then I would have had a clue as to what we were in for. But instead, I made a date for Cheryl and myself to meet with George Tucker and go over a finalized sheaf of forms. Preparing for that sit-down, I dug up the fourteen pages of appendix to Form NRCS-CPA-1202, which had launched our application process. The fine print, so to speak. It made for fairly hysterical reading, although most of the provisos went the government's way. If Cheryl and I had schemed to misrepresent anything, our project would be toast. There were circumstances under which the government could unilaterally change the terms of our agreement, but on our end we were destined to be stuck with it. We had to swear we hadn't been convicted in the past three years of fraud in connection with a government contract. We had to certify compliance with the Drug-Free Workplace Act, and specifically the requirements of something called 7 CFR part 3021; in our younger days this would have surely been a deal-breaker, but we had long since stopped smoking marijuana. We had to certify we hadn't lobbied any member of Congress in order to win this contract—no, wait. We only had to certify that in case the value of the contract ran into six figures. Two thousand dollars was chump change, after all, so we could have lobbied to our hearts' content to get it.

Basically, after a cursory reading I didn't see a clause I didn't feel we could live with. So on July 22, 2010—a mere six weeks since we had first heard of WHIP—Cheryl and I drove to the county agricultural office to go through the formalities with George Tucker. We sat down in what looked like a comfortable board room. Leather chairs, long thick table made of polished wood. A sheaf of nifty documents—maps, schedules, plan directives—had been collected in a fancy-looking binder; the fourteen-page appendix was in there, too, and now George directed our attention to a passage on page nine that I had overlooked. "Recovery of Cost," it read. The gist of it was that right then and there—before I had even made a start at working in the woods—the taxpayers' money had been lavished on our behalf to come up with this thoughtful and carefully detailed plan. So if we signed on the dotted line and then crapped out at some point down the road, the government would be coming after us for restitution. Giving back the money they had spent on us, and then some. Here's how section 12A of the Appendix put it:

> . . . in addition to the refund of payments as set forth in Paragraph 11 . . . the Participant agrees to pay liquidated damages up to an amount equal to 20 percent of the total financial assistance obligated to the Participant in the Contract, at the time of termination. This liquidated damages payment is for recovery of administrative costs and technical services and is not a penalty.

I did some quick thinking. Five hundred bucks we'd have to shell out, if we broke the contract. Plus returning any payments that might have been received. "That sounds kind of like a penalty to me," I said.

"No, it's not. We've already spent that much on you—or probably more, in your case. So we'd need to get that back."

Sure, I thought. All that GPS work, all the mapping, and the multiple site visits. All those miles on government pickup trucks. All that allocation of professional time. But I could think of reasons why our bat contract might eventually head south. Or not see a full and final state of completion—at least not in the NRCS's eyes. They were, after all, a souped-up version of the Soil Conservation Service. I could well recall how hard it was to please *their* people, thirty-odd years ago. Still, there it was in official-looking type: a schedule of payments to be made to Cheryl and me in each of the next three years for working in the woods. Our woods, after all. How many forest owners get to make a deal like that? I looked at Cheryl and we both began to nod our heads. George Tucker reached into his pocket for a pen.

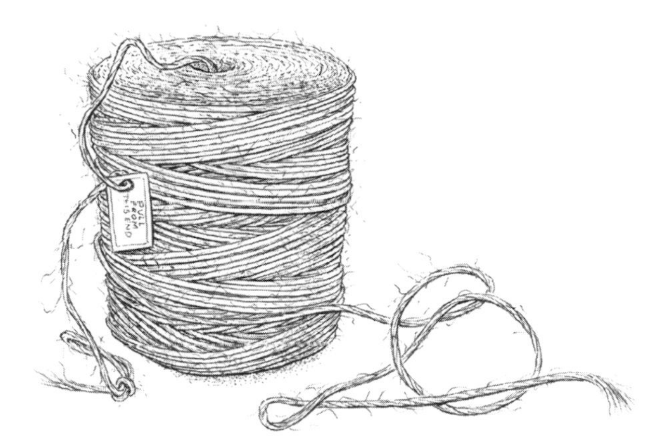

Yahweh's Smile

About a week later, I grabbed a roll of garbage bags and carried them into the woods. It was time to take a stab at pulling garlic mustard. I chose a spot that looked roughly equidistant from each of three shagbark hickories, and knelt down on the ground—what I later came to call assuming the position. My goal was to hit it hard for two or three hours, then measure how many square feet I had covered. From that, I could estimate how many hours—or days, or weeks, or months—of this activity I had in store. This stupid, ridiculous, mind-numbing activity.

My hands got right to work. At first, I was going after both iterations of *Alliaria petiolata*: the first-year rosettes growing close to the ground, and the second-year stalks with their profusion of seed pods that looked to have been triggered by an epileptic fit. But then I found the work went better if I specialized. It was hard to study the horizon of the forest floor at two heights simultaneously; then, too, the technique of pulling out the first-year plants was somewhat different from uprooting the second-year variety—somewhat easier, too. Given that I had a choice, it seemed smart to go after the first-year plants, since getting rid of them would prevent the next year's seed crop. As for the plants that had already procreated, it was late enough in the growing season that many of their seeds must have fallen to the ground. That was regrettable, but nothing could be done about it. At this point, I'd get more bang for the government's buck by focusing on first-year plants.

Working on my hands and knees, I hadn't gotten far before I bumped into the trunk of a fairly large deadfall. Hemlock, by the look of it. The stubs of broken branches held the trunk up off the ground, but they also formed a picket fence that made it hard to get in under there and weed. I could see several green rosettes of garlic mustard—actually, a lot of them—but the plants could not be reached unless you were a leprechaun. I changed course and started crawling backwards; soon enough I ran into a tangle of fallen limbs. There was garlic mustard growing underneath them, too. After only half an hour, I was getting frustrated. But it seemed the task would go more smoothly if I spent some time performing a thorough cleanup of the area: piling the smaller branches into coherent stacks, then cutting up the fallen tree trunks into stove-length chunks. To be used as firewood, after I had built a trail in to let me haul them out.

The site where I'd begun this work was scarcely a couple hundred yards from our house, but I didn't want to waste time and energy trudging back and forth. Particularly with a set of cumbersome tools: pruning loppers, chainsaw, hatchet, sledge hammer, and wedges. Not to mention gasoline to keep the saw running, and a jug of oil to lubricate its bar and chain. I could have loaded all this gear onto the pickup truck, but that seemed like overkill. Then I realized that everything would fit onto the golf cart we'd acquired a few years back. First we had borrowed one at the time our daughter, Anaïs, got married on the farm; the ceremony took place at the edge of the pond, but the party and the feast were in a tent behind our house. A long way to make the older relatives walk. Impressed as we were with the golf cart's ability to whisk senior citizens all over the farm, Cheryl and her brother went out and bought a used one for their parents soon after the wedding. It hadn't been driven all that much, but from time to time it came in handy. Now I decided it could function as my woodsmobile, ferrying me and a variety of tools back and forth from house to job site. Well, not quite to the job site, since I couldn't drive the golf cart in the woods; it had pygmy tires and minimal ground clearance. But it would get me to the edge of the woods, at least. I may have been the only logger in Vermont whose means of transportation was a second-hand EZ-GO.

Chainsaws have only been around for roughly eighty years, but it's hard to think how people logged the woods without them. True, they're very noisy and pollute the pristine forest air, but they're also incredibly efficient—nothing like an ax, which must have been sheer torture for a man to swing all day. When a saw is properly tuned and sharp and running well, it becomes a virtual extension of the user's arms. Once you learn a few basic principles and safety rules, it's possible to wield a saw with great precision and what feels like intuitive control. You aim the business end at the piece of wood you want to cut, and *shazam!* The deed is done. There are many ways to gain mechanical advantage so the saw itself does most of the work, rather than your muscles. And though it is possible to hurt yourself—and badly, too—after some experience the fear of impending doom ceases to cross your mind. Any more than working with a sharp knife in the kitchen.

When a chainsaw starts acting up, on the other hand, you spend as much time making adjustments and repairs to it as you do cutting wood. That was what I found myself doing that afternoon, setting out to clean up a modest portion of the woods in order to do a better job at pulling garlic mustard. First the chain heated up and sagged from the bar, thanks to egregious expansion of its links. I allowed the saw to cool down, then turned a hidden screw that lets you tighten up the chain. Next time it heated up, the chain jumped right off the bar. This is not supposed to happen, but at times it can and does; that's why the body of the saw has a "chain catcher," which you hope to God will work the way it's supposed to. And it did that day, for me. Still, it was disquieting. I got the chain fitted back into its groove and readjusted things; then I gave the saw ample time to cool down again. When it felt good to go, I started up the engine and worked for maybe ten minutes before I could feel heat radiating off the bar. Intense heat. I could smell it. It was hot to touch—a fact that I established by foolishly touching it. Bad idea. The bar left a blister on my thumb.

It was a reputable make of saw—a Stihl—but by the summer of 2010 it was far from new. In fact, I had bought it back in 1972 as an eager young hippie going Back to the Land. For its first decade or so, the saw got lots of use. And though it had languished through my years in academia, there were still occasions when I'd bring it out and fire it up. Every spring a few trees would be found to have fallen, over the course of winter, into open fields; these needed removal before haying could begin. Sometimes a tree would even fall across the driveway, requiring immediate attention before cars could get in or out. The saw did gather dust for months at a time, but I kept it tuned and gassed up—ready for action. Actually, a few years back I'd spent two hundred dollars on chainsaw repairs—more than the machine had cost me when I bought it new. But there were still some unaddressed issues, and now one of them was staring me in the face: the system for delivering oil to the bar and chain was just about kaput.

Andreas Stihl deserves primary credit for invention of the gasoline-powered chainsaw, circa 1929. The company he founded has done very, very well—in part by steadily adding innovations. One of these goes by the registered trade name of "Oilomatic" lubrication. Unlike some other saws, a Stihl automatically squirts a fine stream of oil onto the cutting chain as it flies around the drive sprocket; this in turn lubricates the groove of the bar and keeps a check on heat and friction. On other makes of saw that do not have this feature, the operator actuates a manual device to keep the chain lubricated. No thanks—not while I am trying to stay focused on the cutting job at hand. But my Oilomatic was no longer pumping oil—or not enough of it—and on this vintage model I could not adjust the flow. At the local dealership, I showed my problem to a mechanic who was younger than the saw by several years. I figured that in round numbers half the history of chainsaw manufacturing had taken place since I had bought it. The repair man told me that for all intents and purposes my chain was running dry. No wonder, then, that it kept heating up and causing trouble. Fixing it would run into a fair amount of money, and I'd still be working with a very old machine. He said I should give some thought to purchasing a new one.

I went home and pondered that. Looking ahead to all the work I'd planned on doing in the woods, I knew I had to have a chainsaw I could live with—figuratively and literally, too. And I could afford a new one, since I'd be collecting all that money from the government. So over the next few days I checked out some other brands, shopping from store to store. Husqvarna. Homelite. Echo. All of them had selling points and satisfied customers; why not give Stihl's competitors a chance? In the end, though, I wound up purchasing another Stihl. The first had served me well, and I had come to understand its ways. Chainsaws require almost daily attention, and I'd learned the things that needed checking on a Stihl. Where to find the air filter to brush its surface clean of dust. How to adjust the chain, and keep its teeth filed sharp. How to set the choke for easy starts at any temperature. Lots of things I'd have to learn from scratch on a different make. My new saw set me back six hundred dollars, and it wasn't any bigger than the one that I'd retired. But it was held to be a tool for professionals, not a cheaper "Farm Boss" or a toy for weekend warriors. And it did come with an adjustable screw that governed the Oilomatic system's rate of flow. Also a Rollomatic bar with a nifty wheel that whisked the moving chain smoothly around the nose. It had rubber shock-mounts to absorb vibration, and a braking mechanism that would stop the chain dead in the event of a kickback. It was, all in all, a sweet machine—and now I had a saw that I could use with conviction that the chain wouldn't jump off the bar and try to kill me.

Safety first—yes. My motto. When I bought my first Stihl, the manual they gave me had been fairly brief and to the point. There were standard warnings, sure, but the gist of them was to be careful or you'd hurt yourself. Who didn't know that? Most of the booklet dealt with maintenance procedures and the numbers for replacement parts. The new saw, however, had a sixty-page manual and the first thirty pages laid out eighty-seven warnings about things that could go wrong. Bad things. Scary things. *Warning!* read heading after bold-type heading, with a cautionary symbol to drive the point home. Some of the scenarios were downright gruesome. There was a graphic illustration of a saw kicking back into the sawyer's head. Tiny sparks—or stars, perhaps?—were bouncing off the brim of his industrial hard hat. Luckily, he was wearing serious headgear. But *I* like to work in the woods in a baseball cap, mainly to keep the hair out of my eyes. That, and a good pair of hearing protectors so I won't go deaf in twenty years. I wasn't playing safe, according to the manual. In fact, I was likely throwing caution to the winds. But playing safe was hard to do, given eighty-seven different tragic scenarios to bear in mind. The manual from Stihl had gotten seriously lawyered-up. Once I'd read the whole thing, I was shaking. Shaking in my sneakers, I should note—not the steel-toed safety boots the booklet recommended. Starting up the new saw seemed tantamount to suicide.

Once I got into the woods and put the saw to work, though, confidence returned. I limbed a couple dozen branches off the fallen hemlock, then started cutting up its stem into stove-length chunks. The whirring blade buzzed through the wood with impressive speed; the tool was well-balanced, and the cutting felt effortless. Firewood was piling up. And my mind began to tinker with a bold, ambitious plan—worthy of the six hundred dollars I'd just spent. In a worst-case winter it could take three cords of wood to back up the passive solar system that warms our house. Call it three and a third cords, for good measure. When my father died, he had attained the proud age of ninety-three; I was a couple months away from turning sixty-three. If I were to heat with wood and live as long as long he did, I would need a cool hundred cords to get me by. (A cord is a tightly stacked pile of firewood eight feet long by four feet wide by four feet high.) And it seemed a reasonable goal, now that I'd begun. A nice fringe benefit to cleaning up the woods.

I did some guesstimating as I got to know the saw. Taking the whole of our long-neglected forest acres into account, there must have been twenty cords lying on the ground—reasonably dry and solid, like this fallen hemlock tree. That was a good start on the wood I'd need, right there. As for the standing trees, there had to be at least a cord per acre in misshapen stems. Culls, that is. Not to mention Unacceptable Growing Stock. And each forest acre was supposed to grow half a cord of new wood per year. So the fuel I'd need for thirty years was there, no question. It was just a matter of getting it out.

In my dad's declining years, I knew there was no way he could have picked up a chainsaw and carried it into the woods. But until his last year when the going got tough, I believe he could have kept the fire going in a stove. And he would have tried to, too, whether his doctors felt he ought to or not. So: a hundred cords, and I'd be fixed up for life. Why not? Able to keep warm without giving a hoot about Exxon Mobil, or Halliburton. Or Saudi Arabia.

Where was I going to keep a pile of wood like that? I played with various options as I toiled away. I knew that certain kinds of wood are slow to decompose—and I knew that trunks of trees are more or less round, duh. Lying horizontally, raindrops tend to roll right off them. But it was too much to think that firewood would stay intact for thirty years without a cover. No matter how decay-resistant the species. Still, it could be stacked beneath old sheets of metal roofing. Steel, or aluminum. The panels could be purchased secondhand, since nail holes from prior use would not affect them for this purpose. Used roofing wasn't hard to find, and it was not expensive. Sometimes it was even free. I could use some rocks—or even chunks of firewood itself—to keep the wing-like panels from blowing away. Build a bit of pitch into each stack, for drainage. That might work pretty well, and at a modest cost.

For that matter, I had a fair amount of empty space tucked away in underused barns and sheds around the farm. The sheep flock had been scaled back to just thirty ewes, down from its all-time peak of one hundred. Consequently, our farm's physical plant was not running at capacity. There were lots of places I could stash a few cords of wood. Put it into storage for a decade or two. Or three. Firewood—much like wool—will keep indefinitely underneath a rain-tight roof. So it sounded like a plan. I stopped the saw to fill its reservoirs of gas and oil. It was an exciting afternoon, and I was on my way.

By evening, I paced the area of forest that I'd managed to clean of deadfalls and debris. Sixty feet by seventy feet, which seemed respectable until I calculated that it came to a mere one-tenth of an acre. Or not quite. And I was a hurting cowboy after all that effort. My back ached; my right shoulder tingled, like I'd pinched a nerve. If I worked just as hard for forty more days, I could hope to have the four acres of bat zones free of forest trash. *Then* I could make a real start on the invasives. But it seemed unlikely I could put in forty days like that. Not in a row, at least. I would have to pace myself. Besides, the work could only be done in good weather; rainy days were out. Then, too, we were already into August. The sun was setting noticeably earlier, each evening. Fall was on its way. Soon there would be frosty nights, and days too cold and raw to enjoy being in the woods. There was a finite window in which to get the work done—the work that I had contracted to do for the government. But I'd made a start, at least. I had to feel good about it. Now I had to stay with the program, step by step.

The neighbor's kids from Baltimore arrived a few days later. After letting them settle in, Cheryl paid a visit and explained to them about the job. Since they seemed interested, I arranged a time for them to walk the woods with Ethan and learn to identify the plots of garlic mustard. He was pretty sure that he could make a game of it, sort of like a treasure hunt. I described the area of woods where I had piled brush and cleared away obstructions; that would be a likely place to get the kids started. Give them each a plastic bag, and see who could fill one first.

"Are you going to pay them?" Ethan asked me.

"Sure. I wouldn't ask them to pull weeds for nothing."

"How much?"

I scratched my head. "I don't know what's fair." I didn't want to see a lot of money change hands; I wasn't getting all that much from the government. But if both kids liked the work and kept at it steadily, their wages could start piling up. I felt like bureaucrats from WHIP, trying to assure that scarce resources would be wisely spent. "Five bucks an hour, maybe?"

Ethan frowned. "That's below the minimum wage."

"What is the minimum?"

"Eight dollars and fifteen cents, in Vermont."

"Really?" I was sure he knew, though. He had done his share of working jobs that paid the minimum.

"And it's going up," he said.

But that meant that having two kids working for a day would cost me well over one hundred dollars. And I reckoned there were several days of work to do. "Look," I said to Ethan. "Let them know I'll pay them something, but I don't how much yet."

"You should have a piece rate," he advised quite sensibly. "Maybe like a penny a plant—that wouldn't kill you."

Ah, I thought. Like Japanese beetles in South Jersey, fifty-odd years ago. Given the overall inflation since 1960, one could say the price was right. Still, I had some reservations. "The plants like to grow in family whorls," I pointed out. "Seven or eight together. Sometimes nine or ten."

"So?"

"So pulling out what looks like one plant could cost a dime."

"Start them out at five dollars a bag, then. How about that?"

I was concerned about who would be deciding when a bag was packed full, at five dollars a pop. But Ethan said to try it out and see how things would go. Next morning our young crew showed up bright and early, and Ethan chatted them up as we unrolled plastic bags and headed toward the woods. Hunting garlic mustard to protect the environment—this was an adventure that you couldn't have in Baltimore. If anyone could sell teenagers on pulling weeds, Ethan could. He could get Tom Sawyer to paint a picket fence. So I left the three of them and went off to mind my own business. Twenty minutes later, they came trooping from the woods. "Back so soon?" I asked, dismayed.

"That's not humanly possible," said Alex, the older of the two kids who had given this a go. Matter-of-factly. Telling it like it was. He handed me a bag with several dozen plants tucked in it. Some of them were garlic mustard; some of them were clearly not. "That stuff grows back as fast as we can pick it."

"Everywhere you look, there's more," agreed his sister. "No way."

Ethan was sanguine about their quick retreat; he had done his best, but the game was a hard sell. For starters, it made no good sense to try to rid the woods of something that was growing everywhere. Or nearly everywhere—or so it must have seemed. Heck, it didn't even make a lot of sense to me. I might as well have asked them to pull out all the clover in the lawn. And it's hard to pitch a chore as being a treasure hunt when the treasure isn't all that difficult to find. When it is, for all intents and purposes, ubiquitous. Maybe if I raised my rates to twenty bucks an hour—or forty, or fifty—I could have won their labor. But I doubt it. They saw the task as Sisyphean, and perhaps it was. I hadn't done it long enough to really know, yet.

So the child-labor scheme had managed to crash and burn. That was depressing—and then next day, I checked the mail and found our tax bill from the Town of New Haven. Opening that envelope had always been a pulse-quickening moment of the year, and this time I was hopeful I'd be smiling for a change. We were, after all, in the state's Use Value Appraisal program. Since we had submitted an acceptable Forest Plan and had begun to implement it, other taxpayers would be stepping up to help us out. Our own taxes would be generously abated. That's what I'd been led to believe, at any rate.

No such luck. The benefit of having put most of the farm into Use Value Appraisal for tax relief was around two thousand dollars—or twenty-five percent of what had been our former tax bill. But the new taxes owed for having built a house addition—The Annex, from the year before—came to thirteen hundred dollars and change. Then, too, the town's tax rate had gone up by four percent. All said and done, our tax bill had gone down by a measly four hundred dollars. It was still a whopping sum of money to come up with. And to nail down these miniscule savings, I'd entered in a contract that now had me committed to weeks or months of battling with exotic invasives. *Mano a mano*, the way things were turning out. It wasn't like I could go out and get a job; I was needed right here to crawl around the forest floor with a roll of plastic bags. What I'd gotten into seemed a dubious tradeoff, but the numbers added up: the lion's share of value in the farm was in the house I'd built, rather than the fields and forests surrounding it. Or that's how things looked to the town's team of real estate listers.

One thing was certain: I was going to be needing every nickel I had coming from my contract with the government. Certainly I couldn't run the risk of delinquency and wind up having to pay WHIP for their services. So I returned to the woods with renewed ambition—and a dose of something like adrenaline, too. I was going to tackle those invasives, goddammit. No more procrastination, no more wasting time. I spent most of a day weeding every lick of *Alliaria petiolata* from the plot of forest that I'd cleaned the week before. Two plastic bags full—the thirty-three-gallon size used for lining trash cans. If I saw a plant that bore even a vague resemblance to the dreaded garlic mustard, I would yank it out and bag it. Why think twice? If it had a heart-shaped leaf, a plant was likely toast. By evening, I had that tenth of an acre looking spic and span—as far as that particular invasive was concerned. There was also glossy buckthorn, but I couldn't make a start on that till after Labor Day. Else a glyphosate treatment might not kill the roots. Still, I'd done a good day's work.

But then, nursing a gin and tonic heavy on the gin, I checked some revised maps tucked in with the WHIP docs. Based on the designated zones for bat management—which in turn were based on careful GPS coordinates—it struck me as possible that some of the work I'd done was not inside a designated bat zone. More than just possible. Very, very likely. I felt like shedding tears, or vomiting. Or both.

Next day, I talked on the phone with Brendan Weiner and came away relieved. He said that as long as there were shagbarks near the place where I'd been weeding, I should be okay; I didn't need to get overly stressed about whether I was inside a line drawn on a satellite map by a computer. There was a degree of flexibility here. When the time came to evaluate my first year's work, the government would calculate the area I'd actually weeded. If it turned out to be more than four acres, they could alter the terms of our agreement to reflect the facts on the ground. Before I went much further, though, it might be smart to physically demarcate those portions of the woods where I proposed to do my weeding. Then I could see exactly what was in and what was out—and show it to George Tucker when he came to inspect my work.

Physically demarcate. "How do I do that?" I asked.

"Get a roll of flagging tape. Tear off short lengths and tie them onto tree branches, right around eye-level. Every fifty feet or so."

"That's enough to mark the lines?"

"Sure."

So I went shopping for a roll of flagging tape. It came in various psychedelic hues; I chose an eye-popping shade of orange. Day-Glo. Flagging tape is handy stuff, as I soon found out. You can tear it off the roll with just your bare hands, and a simple overhand knot will tie it where you want. Once in place, the tape is apt to hold up for many years. The main challenge is to not forget why you put it there.

But I didn't like the imprecision of using bits of tape—tied to random branches of occasional trees—to suggest a line I thought should be clear and unambiguous. Beyond any room for interpretation or dispute. So I wound up using my orange tape for a different purpose: I ran a loop of it around the trunk of every shagbark hickory to be daylighted, right at breast height. Sixty-seven trees, in all. That took most of a day to accomplish, but it seemed a job worth doing. Now I—or anyone else, for that matter—could see the target trees from a considerable distance. Knowing where they were became a guide to laying out where I would run a more substantial line, marking off each bat zone. Then I would know exactly where to fight invasives.

That night, I had an inspired thought as to how I'd demarcate those zones: I could use long runs of leftover baling twine. Twine for a hay baler comes in a continuous spool of nine thousand feet, or nearly two miles; the baler automatically cuts the twine to length and ties a couple knots in it every time it makes a bale. Tying these knots only takes a split second. You load spools of twine into a bin adjacent to the knotting mechanism; four spools of twine can be carried at once, enough to put up a thousand bales of hay. But at summer's end, there is always extra twine left sitting in the baler—sometimes an entire spool of it, or nearly. Once it's been removed from its moisture-proof packaging, twine doesn't hold up well. Twine made from fibers of the sisal plant—the kind we use—starts to break down and degrade pretty quickly. If a spool of twine acquires weak spots here and there, there's no way to bale with it. It keeps breaking from the strains imposed by the machine, causing lots of downtime. Consequently, everyone who spends their summers making hay has a good supply of extra twine when the season's done. It sits deteriorating till you take it to the dump. I dug out a half-used spool—four thousand feet, at least—and carried it into the woods; I figured that by walking from tree to tree and wrapping the twine around their trunks with a do-si-do, I could run a line that was precise and continuous. The twine would do its job of demarcating bat zones, then disappear by itself within a couple of years. Thanks to the miracle of decomposition.

Most of the zones as projected on the WHIP map were half-acre circles, so I had to run the twine in a way that would suggest those. I'd choose one particular shagbark as the center of each putative circle, then measure radii from that tree to various points on an imagined circumference. Compass points, as it were. Marked with loops of flagging tape. Eighty-three feet was the radius I calculated for a circle that would enclose half an acre. So I called that thirty paces—maybe on the generous side, but that meant I'd be playing safe. When I had eight to ten points flagged with tape along a circle's perimeter, I'd take the spool of baling twine and make my way from point to point. When I got back to where I'd started, I would cut the twine and tie it off; now I had a shape that was approximately circular, and likely very close to the specified half-acre. Five of those circles, in all. It took a full day's work.

The long, narrow, serpentine management zone enclosing one and a half acres was harder to demarcate, since it was a highly irregular polygon. I wound up having Brendan stop by with his GPS, and we worked back through the list of coordinates taken by the WHIP team. Brendan said that I could afford to be generous here; if the area enclosed came in somewhat bigger than the contract had allowed for, the government would pay for the additional square footage—as long as I performed the work to their satisfaction. So we made the serpentine enclosure pretty fat in places. From end to end it must have spanned a thousand feet, and in certain places it was fifty yards wide. Big, big area. I had barely made a start on pulling garlic mustard, but once the baling twine was all laid out, the scale of our project had increased by another acre. Five acres, then—or four point nine, to be exact. All enclosed by long runs of sisal baling twine that would otherwise have wound up rotting in a landfill.

At this point there was nothing left to do but hit the weeding work, and try to hit it hard. Based on my previous afternoon of piling dead branches and cutting up fallen trees, I figured it should take about two eight-hour days to clean up a half-acre bat zone in the woods. Then maybe two more days to root out all the garlic mustard. And I planned to do the work circle by circle, getting one management zone the way I wanted it before moving to the next. That way, I would give myself a sense of steady progress. And the work would vary, too, from one day to the next. After a tiring day of wielding the chainsaw, I could spend the next day piling up slash and branches. Followed by a day or two of crawling on my hands and knees, filling plastic bags with first-year specimens of garlic mustard. The different tasks had different psychological tenors, too. Working with the chainsaw required concentration; I couldn't let my thoughts go wandering off. Piling up limbs and assorted debris, however, was a different story. And when it came to pulling garlic mustard from the ground—well. Without a lively train of thought to chew on, the work would have very quickly bored me to death.

The winter before, when I had built a house addition with the conscious goal of trying to understand my father—trying to comprehend his ways, if that were possible—the mental project failed because the work at hand required too much constant focus. Now, though, I was doing work that took, at times, no conscious thought at all. None whatsoever. When you spend hour after hour crawling through the woods, studying the forest floor in intimate detail and removing a particular plant that shouldn't be there, the work becomes second nature. Also, it becomes so mechanical, so stupid, that your conscious mind is free to travel anywhere it wants to go. Or anywhere it needs to.

At first, what I wanted to explore was my resentment. I had worked in Hollywood at the age of twenty-two. I had written books and screenplays. I had been appointed to the faculty of an esteemed liberal arts college, opening the minds of kids to literature and film. I had designed and built houses and barns and sheds, graceful gazebos and Thoreauvian *poustinias*—fourteen buildings or major additions, adding it all up. I had helped to raise two children, both of whom had turned out well. More than well. I had presided over reasonably happy lives for five thousand baby lambs, and kept the flock of ewes who raised them well fed for thirty years. And now I was inching through the woods on my hands and knees, working for the government at what seemed like a slave wage. I had transformed myself into a serf. Maybe there was light at the end of this tunnel; maybe my efforts would eventually help the bats. But for the present I kept feeling like an idiot—and like a man on a downward trajectory.

I felt like a monk, too. Or perhaps a novice in some zany religious sect. In Japan, a few years back, Cheryl and I had visited the temple of a cult where devotees spent their days wiping the floor. They had a lot of floors, made of stone and tile and wood. Acres and acres of floors, since this particular temple was their Mother Church. Their Vatican. Grown men and women crawled like slugs across those polished floors, buffing every inch of them with white linen cloths. It was a way for them to humble themselves—I got it. Literally prostrated. I was being humbled, too, although without the blessings of an organized religion. I was being prostrated by a deal I'd made with WHIP. And by a force of nature that had clearly run amok: this crazy plant's determination to overtake my woods.

In my own faith tradition—long since rejected, yet still part of who I was—there was a perception that Nature could eat your lunch. Or that it would try to if you weren't right on your game. Like when Yahweh had his second thoughts about creation and got the bright idea to make it rain for forty days. And forty nights, too. That must have humbled all the people who were washed away. Or like the ten plagues that Moses sicced on Egypt—plagues of nature, most of them. Plague of Frogs, Plague of Lice, Plague of Boils, Plague of Locusts. That must have humbled a lot of proud Egyptians. Now I was caught up in a natural plague, myself. And at times it seemed to be of Biblical proportions. I had been afflicted with a Plague of Garlic Mustard.

Teaching environmental literature to college students, I liked to start the course by asking them to read the first four chapters of Genesis—and read it as though they'd never read the text before. (I was always stunned by how many students never *had* read the text before; Sunday school classes must be going out of fashion.) When it comes to understanding how we see ourselves within the natural world, though—by "we" I mean anyone touched by the broad brush of Judeo-Christian thought—the opening to Genesis is fairly indispensable. And the text prefigures our enduring schizophrenia toward what we call "Nature," inasmuch as it presents not one but *two* versions of our culture's creation myth. In the first version we are gardeners of nature, charged with the role of dressing and keeping it; in version two, nature is under our thumb. Now I had to ask myself which of these paradigms I was appropriating, weeding in the woods this way. Gardeners *do* pull weeds, from time to time—besides their kinder, gentler tasks of cultivation. But do gardeners weed with hopes of wiping out a species? That sounded more like dominion to me—and *that* was a take on the human-nature interface that I disagreed with.

My father had no qualms about dominion, though. If he had been working with me, filling up a plastic bag, there would have been a look of triumph on his face. Scooping up flies with a quick hand and crushing them had always brought him special joy; if we had flies in our house when he was visiting, it was his constant sport. When he kept a hive of bees, he went through excruciating pain to impress on them that he was the boss. Nature was a contest that he liked to feel he'd won. I remembered times when he had gotten badly sunburned—burned till his back had turned bright red, lobster red—and then borne the insult without complaining. As if to show that he was strong enough to take it. (How it must have galled him to be diagnosed with melanoma, long after having won these contests with the sun.) He was like an Indian, seemingly immune to pain and critical of those who weren't. *Mind over matter* was his take on life, his mantra. That was how one proved oneself in an arena in which nature represented an opposing force. The enemy. Victory over nature was the proof of one's exalted status as a human being. Proof of one's dominion.

If you buy into the concept of dominion—the idea that humans are supposed to be in charge of things, and are justified in using nature as they will—then you can behave like Yahweh. You can be capricious; you can blow first hot, then cold. You can carry on like a manic-depressive. You can be incomprehensible and get away with it. You can tell Abraham to sacrifice Isaac, then bring out a ram for him to kill at the last second. Just kidding, Abraham. Or just testing you. If you see Yahweh as a Heavenly Father, and if you're the earthly father to your own adoring kids, maybe it makes sense to you to act in ways like He did. Maybe your behavior gets modeled on His behavior in what you regard as scripture. Maybe your adherence to a Bible-based Christian faith even makes you *want* to do that. Gives you a kind of license. I was on a limb, I knew—following a train of thought while pulling garlic mustard like a robot on steroids. But there seemed to be a grain of truth to my reflections. I was engaged in a project that my father would have tackled in a different way, and with a different attitude. One seeking to confirm his own superiority.

The garlic mustard seemed much thicker than it had before. And it *was*, in certain places. Much, much thicker. Elsewhere, though, considerable stretches of the woods had scarcely been affected. They still had to be checked out, though, inch by inch. Within the circular management zones, I learned to start right at the center of things and then work outward in a widening spiral. There were certain areas where I could move fairly quickly—even getting up onto my feet, from time to time, and walking several paces. But in other places I would look around and realize I'd have to spend the next hour of my life right there. Sometimes in an area no bigger than a bed. There was no pattern to where the weed had taken hold, no rhyme or reason to it. But where garlic mustard seeds had rained down for several years, the forest floor was carpeted with plants that had to be removed.

I thought again about my father's special smile as he'd catch and crush a hapless fly. Plucked deftly from the air. The smile I imagined he'd be wearing if he were beside me, pulling garlic mustard from the damp, fragrant ground. When and where else, I wondered, had I seen him smile that way? Then it came to me: he'd do that when he took you out to eat and swiped your French fries. It was a recurring drama, played out countless times. My father had his own fries, but he wanted to eat mine instead. If I caught him in the act, he'd flash that triumphant grin with just a trace of naked, unabashed cupidity; if he got away with it, his face would stay impassive.

I had been thoroughly trained, as a child, to know that taking things that were not my own was wrong. Was, in fact, a sin. My dad had been instrumental in that moral education. So what was he telling me, by reaching out to steal my food? Either that the fries belonged to him—which they did, sort of, since he would be paying for them—or that the rules he'd taught me didn't apply to him. I would not have dared to reach a hand over and grab *his* fries; that hand would likely have been swatted like a fly. Now, a lifetime later, I could close my eyes and still see that smile on his face. With a look that said, in effect: "See what I can do?" That's how I learned to guard whatever food was on my plate—even in contexts that did not involve food. Or eating.

I felt sure these episodes had punched my buttons, as a kid. If not, why would I be dredging them up now? But it *really* ticked me off when he pulled the same stunt on my own son, thirty years later. It was in a restaurant, of course—a public venue where it wasn't "right" to make a scene. And of course my father knew that he'd be picking up the check. But Ethan—was he four, or five?—began to get upset when he saw his fries were disappearing. His upper lip began to tremble, and his eyes grew wet. And then I got angry, too. So did Cheryl, though she seldom shows offense in public. My father flashed that smile of triumph—a smile of dominion, I am trying to say—and he kept right at it. Stealing French fries even after Ethan was in tears. *See what I can get away with?* asked my father's teasing eyes. *I can do whatever I want. And you can't stop me.*

I didn't want to be like that—to share that attitude—even in this brutal interaction with an unloved plant. On the other hand, I wasn't moved to apologize to each next whorl as I ripped it from the ground. If I did my job well, there would be none left—but I didn't have to be malicious about it. I could assert control, I hoped, without becoming a controlling sort of person. I *wanted* to assert control, since otherwise the men from the government would be unhappy. That was apt to cost me some money I'd been counting on. So was I a gardener, in the metaphor of Genesis? No. I didn't feel that way. But neither did I feel like a dominion freak. I was just a nice guy bent on genocide, insofar as garlic mustard was concerned.

When it was explained to me that some of the affected bats—maybe just a few of them, but some at any rate—might prove resistant to the white-nose syndrome, a certain faith in evolution's methods was restored. It synched up with what I'd learned by talking with Kirk Webster, a nationally known and controversial apiarist who keeps a couple dozen hives on our farm. A couple dozen hives, I should say, out of approximately nine hundred. Kirk doesn't go in for artificial remedies to spare his hives the risk of going down from one thing or another; he accepts his losses as part of a cycle that results, eventually, in stronger hives and better bees. Thanks to this philosophy, he's able to sell queens all across North America with the expectation that they—and their progeny, too—should be equipped to fight back against whatever ails the bees. Varroa mites, or foulbrood, or Colony Collapse Disorder—whatever. If I bought into this line of thinking, the bats were going to be okay after suffering a period of losses. Tough losses, to be sure. Egregious losses. The bats-as-individuals were certainly at risk, but a bat *species* ought to demonstrate resilience. Sooner or later, an immunity would be revealed. Some cohort of bats would prove resistant to the white-nose syndrome—hopefully enough of them to maintain a population. Some cohort should likely withstand *any* biologic challenge. Because the genes of individuals are not the same. All eggs haven't been placed in a single basket. It was like ships running a naval blockade: if one succeeds in getting through, its cargo can make up for several other ships whose fate was to be sunk at sea.

But what about a human who sets out to pull up all the garlic mustard in his woods, based on distinguishing physical traits? This was not like white-nose syndrome, or like mites infecting bees; I was damn well able to root out every one of them, as long as the plant didn't mutate into something that I wouldn't be able to recognize as garlic mustard. I was getting pretty good at making those discriminations. White flowers, heart-shaped leaf, growing in rosettes—poof. Gone, Jack. In the bag. Which meant that I was working outside evolution's playbook. Yes, it was a genocidal act I was committing. I was not an ordinary biologic challenge; I was out to kill them all. And with time enough, I'd do it.

Only in the designated management zones, though—that thought came to mind a couple minutes later. Step outside the lines that I had marked with runs of baling twine, and the garlic mustard was alive and well. Thriving, even. That's why it was destined to succeed, in the long run. I was nuts enough to take on five acres of invasives, but that was a drop in the proverbial bucket. After two years of following the WHIP directives, I wasn't likely to be doing this again. Life was too short for this—my life, anyway. The garlic mustard would continue doing its thing, and eventually reinfest the zones that I had rid of it. So what I was doing was essentially absurd—and would be even more so if I did it in my father's head. His take-charge attitude, his need to control things. Try to control a plant like this? Fat chance.

Some mornings, Cheryl would make time to spend an hour or so helping me to weed the woods. We'd be working well within earshot of each other, but without much conversation. Her back would start to hurt her if she stayed on hands and knees, so more often she'd be lying on her side. Like a swimmer, I thought, moving forward at a glacial pace. Two people could be more than twice as productive as one, thanks to shared effort. I valued my wife's help, and the warmth of her companionship. Crazy as it was, we had a sense of camaraderie while working at this stupid task. Back when we'd been hitchhiking together in the sixties, we'd get into fixes that were equally improbable. We'd have to spend the night in a concrete drainage culvert, or in a ditch along the side of the road. Once we had to wait for seven hours to get a ride—seven hours of doing nothing, but it hardly mattered. Time imposed no pressure on us, since we loved to be together. That's the way it felt again, those mornings in the woods.

One night I woke in my sleep scratching a clot of blood that seemed to be protruding from my upper thigh, beneath a buttock. That wasn't right, I knew. I shook Cheryl awake and switched the light on; sure enough, a fat tick was feeding on my substance. I couldn't see the thing because of its location, but Cheryl found tweezers and proceeded to drag it out. First the swollen body, then its pointy little head. Everyone who lives in New England knows about the risks of Lyme disease; Cheryl has a couple of friends whose lives have been upended by an unattended tick bite. So in the morning I skipped going to the woods and went to see my doctor instead. She checked my thigh for a telltale bull's-eye rash, but couldn't find one. So far, so good. There was like a morning-after pill to handle tick bites, and she gave me a prescription for one of those. Seventeen dollars—ouch. But it put me in the clear. Still, I could not prevent repeated risk of tick bites, given the way that I was spending my days. Are ticks an invasive species? I couldn't help but wonder. What about Lyme disease?

That afternoon, I was back in the woods—and conjuring my dad again. There was another time when he used to flash that smile. Yahweh's smile, it suddenly occurred to me. At family celebrations—like Thanksgiving or Christmas—he would organize a parlor game called "Who, Sir? Me, Sir?" After his death, we discovered this had been a drinking game he'd learned in college; that was kind of shocking, since as a grown-up he was sober to a fault. To play "Who, Sir? Me, Sir?" people sat in a circle and each one was assigned a number based on their seat. One, two, three, four—right around the room. Then my dad would say, in an incantatory singsong: "The Prince of Paris lost his hat. Who stole it? Number *four*, number *four*, number *four to the Foot!*" *The Foot* was the last seat in however many seats there were, and it bore the highest number. When my father sent a hapless player *to the Foot*, everyone beneath that person—lower on the totem pole—got to move up a notch. They moved to the next higher seat, and took that new seat's number.

Going *to the Foot* was like going to the dunce's chair. Going *to the Foot* was like being damned to hell. But if number four (or whatever number had been called) could say "Who, Sir? Me, Sir?" before my father could say "*to the Foot!*" then that player had managed to beat the rap. He didn't have to go *to the Foot*. He could go on sitting where he was, and keep the same number. Then my father would commence a ritual conversation. "Yes, Sir. You, Sir," he'd say in a stern voice. As if you were likely guilty, but he'd deigned to hear you out. "No, Sir. Not I, Sir." "Who, Sir, then, Sir?" "Number ___, Sir"—and number four would call out some other player's number, putting that person to the test. Let's say number four accused number three of thievery, stealing the Prince of Paris' hat; number three would have to call out "Who, Sir? Me, Sir?" before my father could say "Number three *to the Foot!*" If there was a close call in the timing of responses, my father was the arbiter of who had won. Often he would favor himself, but not consistently. His judgments could be patently unfair, but they could not be challenged. *He* could not be challenged. And every time he judged that someone had to go *to the Foot*, there was that smile of self-satisfied authority Like he'd crushed a bug, or pulled a weed. Or swiped a French fry.

So working in the forest was a time of revelation. I had many hours to think—many, many, *many* hours—and no clear agenda for the path my thoughts were wont to take. Under such circumstances, what makes a person think of one thing but not another? There seemed to be a sixth sense involved, like when a bat flies by echolocation. What were the insects I was trying to run down, and what were the pitfalls I was trying to avoid? Memories of my father are what I kept coming back to. Some were pleasant, some were not. But it seemed that, day by day, I kept drawing sharper distinctions between me and him. My way of being in the world was not like his, and it didn't have to be. Still, I couldn't help but admit that I was working with a thoroughness—a stubbornness, too—that could only have been learned at my father's knee. He was the only other person I had ever known who would have taken on such a crazy task as weeding five wooded acres of a tiny plant called garlic mustard. And then, having accepted the job, trying to complete it to a standard of perfection. For better or for worse, I was finally taking the measure of my dad. And I badly wanted to. It seemed to be my purpose here.

There were several obstacles that threatened to derail the work, akin to waking up to find a tick burrowed in my skin. For two or three days at the end of August, the woods were lousy with mosquitos; then the wind shifted and they largely disappeared. One day I was working in a far-flung bat zone when my hands began to itch. Then my arms and neck became affected as well. I had handled something that was *not* garlic mustard, and it had triggered an allergic reaction. Back at the house I swabbed myself with Benadryl and napped for several hours; when I woke up, the rash had largely disappeared. As soon as we got into September, there were hunters out. That meant dressing in brightly colored clothes and otherwise trying to make my presence obvious—as obvious as I could make it, crawling on my hands and knees. Little by little, though, the work was getting done. And I was no longer entirely resentful. I was learning certain things and, in an oblique way, almost having an enjoyable time. I had started this project with a mess on my hands—started from square one, from scratch—but most days were marked by a sense of accomplishment. There was slow but steady progress. In the well-marked bat zones, I had garlic mustard on the run.

The point of "Who, Sir? Me, Sir?" was to make your way up the ladder, chair by chair or rung by rung, till you wound up sitting in the number-one position. If you ever got there, then you hung on for dear life; everybody else in the room would be gunning for you. And my dad, especially—setting traps and speaking faster, picking up the pace until he'd finally nail you. When he did, there would be that smile on his face. "Number one *to the Foot!*" he'd shout, and whoever had been first was suddenly in last place. Then you had to start from scratch to claw your way up again.

Were we having fun? I wonder. Fun or not, everybody loved to play that game. No family celebration was quite complete without it. Aunts and uncles, cousins, siblings—everyone on edge and getting pumped with adrenaline. That was what you needed to avoid going *to the Foot*. We were a competitive bunch; we had fast reflexes, hair-trigger response times. Maybe other issues in our family's dynamics were being acted out as we took each other down. Yes—that was almost certainly the case. But in nearly sixty years of playing "Who, Sir? Me, Sir?" my father was the only one who ever got to run the show. Everybody danced to his intimidating tune. And that made him very happy—happy as a man could be, I would almost say. Happy as *he* could be, at least in social settings.

If he had been working with me, pulling garlic mustard out and stuffing it in plastic bags, I could imagine my dad shouting *to the Foot!* again and again. His way of saying *Take that!* to this invasive plant. But there had to be another way to go about the work. Some other attitude, or calculus. Some other mindset. There just had to be. I was not my father—that much I'd decided, and was trying to believe. This was not a contest pitting me against nature, and whatever animus I felt toward garlic mustard was largely artificial. A way to get my tired, aching body out of bed each day and back into the woods. But if I was not playing the role of a dominator—not pulling weeds from a perspective of authority—then how *was* I doing it? And, by the way, *why*? There had to be more to it than fearing the wrath of WHIP and trying to avoid its lash.

By Labor Day—September 6th—there was no more garlic mustard left in the bat zones. Zero, zilch. All bagged up. Now the plan's directives said I ought to buy some glyphosate and go after glossy buckthorn. Japanese barberry and honeysuckle, too. After my initial reservations, I felt more than ready. Several weeks of pulling weeds by hand had slowly worn me down. Now I was up for better living through chemistry. With the federal government and Vermont's chapter of The Nature Conservancy firmly on my side, I wanted to find out what an herbicide could do.

Batshit

At the local hardware store, I pored over a vast array of glyphosate products. Finally I settled on a pint of Roundup Concentrate Plus, thinking that Monsanto ought to know how to make the stuff as well as anyone. *Kills the Roots! Guaranteed! Results in 12 Hours!* promised the label in both English and Spanish. The plastic container came with an instruction booklet, taped on in such a way that you couldn't open it and read the instructions till you'd purchased the goods. Then I noticed that the concentration of glyphosate in Roundup Concentrate Plus was eighteen percent. That was problematic, since the government had specified a twenty-two percent solution for killing buckthorn with their "cut stump" treatment. Concentrates can always be diluted to a lesser strength—in this case, simply by adding water—but there is no easy way to make them stronger than they come. I checked the shelves again; eighteen percent was as strong a glyphosate as the hardware store was stocking. If I was going to go to all this trouble, though, why not do it with a product that would do the job?

I put the pint of Roundup back on the shelf; then, back home, I did some online investigating. Eighteen percent was actually a glyphosate whisky, rather than a beer. For general-purpose weed control, a concentrate was usually diluted till the glyphosate content was at one percent. Even for stubborn weeds like quackgrass and kudzu, a two percent solution was supposed to be effective. So I was being asked to use this herbicide at many times its customary strength. Of course, for general weed control glyphosate compounds are applied as a foliar mist, using spray equipment. Painting the stuff directly onto stumps of trees was not a common application. Still, I had my orders. I knew what I had to do. And it made no sense to do it with a concentration that would not be up to snuff.

A couple days later I was at the local feed store, buying bags of grain to put a "finish" on the three lambs—out of fifty born that year—that were destined for our freezer. When a lamb is nicely finished, its carcass has a thin bark of fat that makes the meat taste even more flavorful. And more tender, too. If you do things right, a chop will melt in your mouth. Feed stores carry all the products that a farmer needs: seed and fertilizer, animal vaccines and brightly colored ear-tags, even watering tanks and wire fencing. So I figured they might have some glyphosate, too. I had never browsed the shelves of chemicals before, but now I did and I was far from disappointed. I found a quart-sized container of something called KleenUp, which turned out to be a high-test glyphosate. Forty-one percent—wow! A double-strength rum. Nineteen dollars and ninety-five cents. The woman at the counter gave an unsolicited review. KleenUp really works, she told me. Maybe not as *fast* as certain other glyphosates, but not to worry. A slow kill was just as good.

I bought the quart of KleenUp and took it home with me—well, not actually into the house but as far as a cluttered workbench in our garage. I poured a few ounces into a Mason jar, admiring its iridescent Kool-Aid hue. A bold, bright magenta. Then I added just about the same amount of water—very slightly less, though—and swirled the jar to mix the two liquids. *Voila*. Just by eyeballing, I had made a concentration that would have to weigh in at twenty-two percent. Maybe one or two percent more, for good measure—but no less. Ballpark.

Then I did some thinking about personal safety. This was not gasoline, and this was not New Jersey in the 1950s. If I were applying

KleenUp on someone else's land, I'd need to have a license from the U.S. government—and I'd only get that license after taking classes and passing a test. This was no joke; this was serious business. I had no thought of stumbling through the woods with an open jar of high-test glyphosate in my hand. So I tucked the jar into a two-gallon plastic pail, packed in with two or three other jars—empties—so it wouldn't slide around in there. Then I found a one-gallon pail that would fit inverted over the mouth of the larger pail. Grabbing the larger pail's handle, I found that I could carry the herbicide without having it slosh or spill. And there was no way that it was going to get on me. Then I found a paintbrush that I'd used the year before to stain some trim boards for The Annex: a two-inch brush that was still in working order, but of no real value. That fit neatly in the mouth of the Mason jar, and its wooden handle didn't poke up any higher than the smaller pail I was using for a cover.

With that crude set-up—and with, of course, the chainsaw—I drove the golf cart to the edge of the woods. The buckthorn was mostly coming up in well-established clumps, and by now I thought I knew where most of them were. According to instructions from Toby Alexander, I had to treat a buckthorn's stump within five minutes of severing its stem; any longer interval would compromise the goal of having herbicide sucked down well into the roots. But it made no sense to have to start the saw—and stop it, too—for each individual specimen of buckthorn. Starting a chainsaw is a muscular exercise; often you have to pull the cord several times before the engine fires up. So I devised an energy-conserving plan: I would cut down ten or so buckthorns at a time, then turn off the saw and get all the stumps painted before five minutes had elapsed. The buckthorns' wood had a sickly yellow color, and its inner bark was jet-black; both of these reacted with the pinkish-purple liquid in a way that seemed to certify the deed had been accomplished. The glyphosate lathered up on contact with the fresh wood, making a satisfying herbicidal froth. And against the kinder, gentler aspects of my nature, when I saw those suds I smiled. Underneath each stump was a thick, tenacious root system, but I had just deployed a weapon that would Kleen it Up.

The history of chemical herbicides is rather brief—and, like the chainsaw, it is hard to comprehend how people got along without them. Basic forms of agriculture date back to at least 10,000 years BC, but 2,4-D—the first real chemical herbicide—didn't make the scene until 1947. That's the year when I was born. The Sherwin-Williams company, better known for making paint, brought 2,4-D to market; later it became a component of Agent Orange, used to defoliate much of Vietnam. The next major broad-spectrum herbicide, atrazine, came into widespread use in the late fifties—right about the time when I was gathering Japanese beetles and drowning them in jars of gasoline. Often used on corn land prior to planting, atrazine has proved to be regrettably persistent in soil and water. That's a major drawback, and it set the stage for Roundup's introduction by Monsanto in 1974. That's the year when Cheryl and I moved to Vermont.

Since then, other herbicides have come onto the market—but a big part of the action in recent years has been the development of genetically modified crops. GM plants have been engineered to *tolerate* particular herbicides, allowing farmers to apply them on their fields even after the desired crop has emerged from the ground. The herbicide then takes out competing weeds but spares the harvest. Monsanto now offers "Roundup Ready" seed for alfalfa, cotton, corn, soybeans and canola, with more in development and soon to hit the market. It's a brave new world out there, down on the farm. But all these herbicidal products are coterminous with my own life—so, not a long history. Sixty-odd years at best. And sixty-odd years is the blink of an eye given the long, slow, tortuous development of agriculture. For the vast preponderance of time that humans have grown food, the only way to tackle weeds was with some form of cultivation. Plowing, say—or hoeing in between a crop's planted rows. Or, as I now well knew, pulling up weeds by hand. There were no significant chemical alternatives. It occurred to me that prior to the very recent past, getting on top of a buckthorn infestation would have been impossible. Which meant that no one would have tried to make me do it.

After cutting down a group of buckthorn trees and painting their stumps with KleenUp, the next chore was dragging all those stems out of the woods. I needed to do this because many of them had ripe berries; inside the berries lurked a host of potent seeds. But the buckthorn stems could be so skanky—so impossibly twisted and intertwined—that it helped to cut them into sections ten or twelve feet long. Once piled up in the field at the forest's edge, I planned to let them dry until conditions were right for burning. Maybe not till winter, when a blanket of snow would prevent the fire from spreading. So after a round of cutting buckthorn stems and painting stumps, I'd have twenty minutes' work to do as a human mule. Stupid work—which meant that I could let my mind go wandering. Sometimes, though, I'd let my head go empty in a zen-like state. And sometimes I would sing a song over and over, till it turned into an earworm and I could not let it go. Physical work and the singing of songs have likely always gone together. *Come mister tally-man, tally me bananas. Weigh, haul away, we'll haul away Joe*. The music of a work song creates a rhythmic context for the physical activity that people are engaged in, and its lyrics tend to reinforce the social value of whatever's being done. Thus the time goes by more quickly—or at least it seems to. Work and song are made for each other, one could even say. Kind of like bats and shagbark hickories.

I was raised in an extraordinarily musical family. Not to the extent of the von Trapps, maybe, but close enough to share their perception that music formed the core of a familial identity. Also like the von Trapps, in my family vocal music—singing—was held to be especially important. Standing around the piano as a family unit, crooning hymns and Christmas carols in four-part harmony. Or driving in the car, singing Stephen Foster songs and musical rounds. My maternal grandmother was a well-known voice teacher in Chicago's fine-arts world; she had a studio on Michigan Avenue. As young kids growing up in a northwest suburb, my sister and I were taken to The Loop each week for singing lessons with this diva. Our cousins from Lakeview took lessons with her, too. Esther Bowker was a strong and charismatic woman, and she was the choir director at our family's Baptist church—a choir in which my mother was the star soprano, and my father a determined tenor. My aunt and uncle sang in that choir, too. By the age of six or seven, Esther Bowker's grandchildren were all accomplished soloists. It was expected of us. It was who we were.

This was all before my father took a job in South Jersey, making us decamp from Chicagoland when I was nine years old. I've never understood his reasons for this bold decision, or how he managed to impose it on my mother; it's fair to say that her extended family felt betrayed. But it's tantalizing that the move took place when I was just the age my dad had been when *his* father died. As if to make me understand that life meant adaptation. Unexpected news could roil a person's life at any time, overturning comfortable patterns of existence. Out of his father's death, my own father seemed to have forged his personality—one based on the pillars of mechanical competence, emotional austerity, and exercise of iron will. He had found out who he was by dealing with a tragedy; now his kids would have the chance to do some learning, too.

Music stayed important in our lives after moving East, but we were no longer pushed to prove ourselves onstage. There were no more voice recitals, no more child solo gigs in front of rapt congregations. Once removed from that expectation of performance, I began to doubt the values that had come along with it. I came to believe that we'd been raised like trained seals, trotted out to bark on command in exchange for love—or at least expressions of affection and self-worth. As if our purpose was to make the grown-ups proud of us. Why couldn't we be loved for who we were? I wondered. Whether we sang well or not. Whether we sang at all. Why should a child be considered a "better" child just for getting up on a stage and performing? Music still provides me with enormous satisfaction; I'm the kind of person who can't get through the day without a song playing on my lips or running through my head. But it's there for my own pleasure, nowadays. And no one else's.

In view of this history, it's wonderfully ironic that my daughter has become one of the most admired singer-songwriters of her generation. Consequently, for the last eight years or so the work songs that I sing to myself are often hers. Given that I had been raised to be "musical" and then made a conscious choice to turn from that path, I never encouraged musicality in my own kids. And since Cheryl can't quite carry a tune, she did not insist that they be musical either. But the genes were lurking there. And when they emerged—when Anaïs, as a high school student, started writing songs and asked if she could take guitar lessons—we did not object. We helped start her down what became a long road, then watched her come into her own and claim her gifts. Now she has a songbook that is thick and artfully eclectic, filled with lyrics that deserve to be approached like poetry. These days, she gets onstage and sings in front of audiences several times a week—and she does it with a grace and ease that never came to me. That's what you get for *not* pushing a child to pursue a latent aptitude.

So that morning, dragging stems of buckthorn to the field after poisoning their stumps, one of Anaïs' songs was playing in my head—and playing there, no doubt, because it had a certain resonance with what I was doing. Human transactions with nature were on my mind, and I was involved in a particularly gruesome one. The song that came to me is called "A Hymn for the Exiled," and its central image is Yahweh's expulsion of Adam from the Garden of Eden. Other forms of exile are alluded to as well: the way that every grown-up is an exile from childhood, the way that lovers who break up are exiled from each other. In some ways, "A Hymn for the Exiled" is a counting song; its opening line, for example, goes "One-two-three-four-five-six-seven." Referencing the time it took for Yahweh to create the world. Counting was appropriate, though, for hauling stems of buckthorn step by step. So was the song's emerging thematic freight: the speaker soon morphs into a spokesman for humanity, daring to chide Yahweh for His banishment of Adam.

Did You see how far he fell?
Did You watch him
Covering his body in his shame?
Wanting You near him
Though You couldn't hear him
While he was falling down
With Your name in his mouth.

Of course, there's something gratifying about having one's own daughter share her dad's concerns. The Old Testament is a text I keep returning to, long after having been made to learn it as a child. I don't return to it as anything like scripture, but as an unwitting record of what humans seem to have thought was going on here. Or what Judeo-Christian humans thought was going on. And a sense of separation from the divine—from the putative source or sources of creation—is part of that record; so is the perception that we used to have a better, more harmonious relation to the natural world. And that we managed to lose that relationship on account of some essential trait within ourselves. How could our Creator have made us the way we are, and then hold that against us? That is the implicit question of the song.

I could relate to Adam's banishment with every step. He had been condemned to a life of struggle versus nature; the ground itself had been cursed by God against him. Now he had to sweat to make it yield plants that he could use, rather than thorns and thistles. Well, now—I was fighting thorns and thistles, too. Driven to extremity, I'd become their murderer with the help of my new Stihl and a jar of KleenUp. But was this the way I would have chosen to relate to nature? Hell, no. I would have preferred to be a gardener. But I'd signed a contract to "control" these invasives, and control turned out to be a code word for domination. It occurred to me that what a person loves, they nurture; what a person fears, they seek to control. Open palm, or closed fist. Two different ways of being, and of interaction. And the way I'd chosen was the way of Adam *after* Eden. In the spirit of the song, it struck me as a cruel injustice. I had a right to be upset with Yahweh, too.

Then something wild happened: all the trees around me started taking on a warm glow. The early autumn air was crisp; the woods seemed shot with sunlight filtered through a scrim of changing leaves. But the colors that I saw were anything but natural. I dropped the stem of buckthorn in my hands and rubbed my eyes; my vision did not clear. I closed my eyes for maybe ten seconds, then reopened them and blinked a couple times. Now the woods had turned magenta. The same hue, in fact, as my solution of glyphosate. Feeling dizzy, I sat down to think the situation through. Had I made some terrible mistake? As far as I could tell, I had not spilled a drop of KleenUp on my clothes or on my shoes—and certainly not on what little skin I'd left exposed. True, I was "supposed" to be wearing latex gloves. And plastic goggles, too—but a person has to draw some limits. I was not going to run a chainsaw wearing latex gloves, and I wasn't going to pull them on and peel them off again every five minutes. I was wearing regular glasses with corrective lenses, adequate to give my eyes basic protection against a splash of chemicals. But there hadn't been one. So had I inhaled the stuff? Probably. But not the way I would have if I'd been behind a spray gun.

So much for maintaining a Drug-Free Workplace, I thought with dark amusement. If I'd taken LSD, I couldn't have expected to see any brighter colors. Then I felt the rumblings of a panic attack—*bad trip! bad trip!*—and thought I'd better drive the golf cart home and put myself to bed. I didn't do that, though. I told myself this had to be a psychological response to breaking with my longstanding prejudice toward herbicides. The warning label had said KleenUp might cause "substantial but temporary eye injury," with attendant distortion of visual faculties. I was pretty sure I hadn't injured my eyes, but I had processed the warning and filed it away in some corner of my mind. So it was my mind, perhaps, that had played this trick on me. Then I thought: "Look—you can give in to anxiety or just play through. Either way, you'll see what happens." I decided to play through, but leave the pail of glyphosate waiting for another day. I went back to dragging buckthorn stems to the nearby field, and after ten minutes I was seeing normally. The woods had ceased to shimmer with a palette of magenta. *So*, I told myself, *chill*. Stop acting crazy. And find some other work to do than painting stumps with KleenUp.

It wasn't hard to find some work that wouldn't freak me out. All around the larger buckthorn specimens were tender shoots, many of them young enough to pull out by hand. Once I got the hang of this, I learned that if I clamped a heavy pliers on the woody stems I could gain more traction; then a plant bigger than my bare hands could ever pull would pop out just the same. Amazingly, the shoots nearly always came out roots and all—and that meant a sure kill, no herbicide required. Stem and roots of buckthorn plants were fused, almost like they had been welded together. So I spent a couple hours doing for the buckthorn what I'd done for garlic mustard plants: mechanical removal. Off in the distance, I could see my plastic pail with its hidden jar of glyphosate. Had I somehow warped my vision by exposure to it? Then I shook my head and thought: "You're a fucking idiot. Come into the woods alone, and mess around with herbicide? Mixed at over twenty times its normal working strength? That's completely batshit. Then you start hallucinating, but don't even go for help? You should have called the doctor, or maybe the poison hotline. You should have had Cheryl take you to the ER. Are these buckthorns really more important than your sight?"

So my inner scold seemed to be alive and kicking. Then I got to thinking about "batshit" behavior: when a person's actions seem *deliberately* irrational, much like the way we tend to regard a bat in flight. Not clinically insane—no, not that at all—but seemingly erratic and unable to make sense of. Hence the expression "having bats in the belfry" as a metaphor for conduct that is aberrant and wayward. The belfry was the human mind, and the bats were—well, the bats were whatever was driving a plan of action chosen in the dark, and chosen on the fly. Moment by moment. It was far removed from ordinary, conscious thought; it was more a consequence of something like echolocation. And it struck me that when a person behaves like that, probably there's something to be learned about their makeup. What had I revealed by staying on the job as the woods around me turned magenta? Right away, I knew it showed I was my father's son.

Once when I was growing up outside Chicago, a swarm of wild bees landed on a tree in our back yard. My father called a local beekeeper and got the man's help moving the colony into a standard wooden hive box. After a month or so, my father opened up the hive. He had made a bee suit out of various odds and ends; it involved a pith helmet and leftover screening from a porch that he had lately built. There were canvas gloves, too, modified by sewing elastic in the wrists. But when my father put this armor to the test, certain flaws revealed themselves. Bees crawled underneath the sleeves of his flannel shirt, then beneath the nylon mesh that draped across his face. They stung him many, many times—they stung him half to death, I'd say. But he didn't quit until he finished what he'd come to do. Checking every wooden frame for comb and brood and honey. Searching till he'd found the queen and saw that she was laying eggs. His bees. His hive. He wanted to show them who the boss was, I expect—who they all were working for.

That night, he came down with a high fever. He became delirious. A doctor was consulted, and my mother ran a cold bath and helped him into it. Next morning, he was on the mend—though not too keen about beekeeping anymore. Still, he had succeeded in getting the job done. Batshit, yes. Not insane, but acting damn crazy from my childish point of view. Probably from an adult perspective, too. It left me with a lasting image of my father tweezing stingers from his swollen arm. One after the other. *Ouch!* Trying not to flinch. *Yow!* Doing his best to keep a straight face. *Damn it all!* And was that the kind of person I wanted to be?

At that point, I gave up on the day's work and hauled my ass out of the woods. Just because my father had to finish anything he started didn't mean that I had to. I had many days of killing buckthorn ahead of me; what was an afternoon of playing hooky, more or less? Back at the house, I didn't tell Cheryl what I'd seen—how the woods had started taking on a psychedelic glow. But I searched a catalog of farm and garden tools, looking for devices that would function like my pliers did but were specifically designed for pulling woody plants. Something that would give me the advantage of leverage. Sure enough, other people had been here before me. There was something called an Extreme Brush Grubber, with spiked jaws that hooked onto the base of a sapling; then you could attach a chain and pull small trees from the ground with a truck or tractor. The tool could be had for one hundred fifty dollars, and a chain of recommended strength would run me sixty more. But I had a lot of work to do in places where there was no way to *get* a truck or tractor, so the Extreme Grubber wouldn't work for me.

Then I started searching "grubbing tools" on the web, and quickly came across something called The Extractigator. Made by a Canadian inventor named Shawn Taylor, it looked like a steel bar attached to ratcheting jaws that could pivot on a fulcrum-like base. The jaws could be clamped onto the stems of woody plants up to two inches thick; when you applied a bit of muscle to the bar, the stems were supposed to pry loose from the soil. Roots and all, the copy read. Even glossy buckthorn. About two thousand Extractigators had been sold, so it was more cottage industry than big business. I liked that. But there were no local dealers, no way to see the tool first and try it out. Buying an Extractigator and having it shipped to Vermont would run a couple hundred dollars, Canadian—and if I didn't like it, sending the tool back to British Columbia would be on me. So I took a pass on that.

Once I had found the Extractigator, though, I quickly found sites for several comparable tools. There was something called a Weed Wrench, made in Oregon. There was a Root Jack—and a Root Talon, too. All of them had something that would latch onto a woody stem, attached to a lever that could yank it from the ground. But now I realized I already owned a tool that could maybe do that; I had a four-foot pry bar with a V-shaped claw. By using a base plate so it wouldn't sink into the ground, I should be equipped to pull out buckthorns that would fit the claw. Up to an inch in diameter, roughly—and that would save me many hours wielding a paintbrush dipped in toxic glyphosate. Any plant that came up by the roots would be a goner.

So that was a good afternoon's investigation. Next day, I was pulling buckthorns with either the pry bar or my heavy-duty pliers. Dozens of them. Scores. Hundreds. I was a happy guy. The bigger stuff could wait till later, once I had processed my episode of seeing things. As for the KleenUp, it could stay right where it was. Where I'd left it sitting at the base of a hickory tree. It was in its plastic pail, covered from the elements. As long as some woodland creature didn't tip the bucket over, it should be fine there. And for the time being, I was going nowhere near it. Hopefully I'd have it all used up before the work was done—hopefully that bucket would never come back from the woods.

Thinking a bit more about my father and his bees, I realized he must have had the worst sort of temperament for working as an apiarist. Certainly he was the opposite of Kirk Webster, who comes over now and then to open his hives here. Once or twice I'd watched him from a comfortable distance. Kirk has an extremely calm demeanor; his blood pressure must be like a hundred over sixty. He works his bees with only minimal protection, and is only rarely stung. My father, on the other hand, liked to make a strong impression on nearly everything he chose to interact with. Anything or anyone. True, his bees had formerly been living in the wild; they had not been cooped up in a hive box before. Maybe that had made them angry when at last they met their keeper. Maybe they had a few complaints for the landlord. But it was a habitat designed to satisfy their needs, with apertures and clearances measured to a bee's dimensions. It was a type of house that other hives had learned to live in, notwithstanding that it *also* was designed to facilitate the theft of honey. Anytime the beekeeper had a mind to do so.

That got me to thinking about artificial *bat* houses. I had come across them when I got into the literature on how to help endangered bats. You could buy a bat house ready-made, or buy a standard kit. You could even build one out of scraps of surplus lumber. The wood on the interior had to be roughened up in order to get bats to roost; the house needed a landing pad and several inner baffles made of plywood or veneer. It had to be dark inside, but set up in a place where the box would catch the sun. That was for warmth, not light. Bats were all about the dark. There were a couple dozen do's and don'ts, based on what biologists had learned the bats preferred. But I got the strong impression that a lot of bat houses failed to win occupants—or failed to be occupied by bats, at any rate. This did not surprise me, though. Artificial habitats can rarely match the real thing.

But artificial habitats do at times reveal what is on a creature's mind. When I built our garage, a particular framing detail at the gable ends soon enticed some barn swallows to move in. Then more kept arriving; it was like that Hitchcock film. Once they'd built their nests inside these long, narrow apertures, the birds would dive-bomb anyone who heedlessly encroached on what they took to be their space. Then, too, the car and truck became awash in bird poop. Finally I spent a day plugging the specific holes the swallows had been drawn to—a cavity created where each purlin of the roof deck lapped across an outer truss. Within a day, all the barn swallows were gone. There were other, comparable pockets they had access to in other corners of the garage. But these were not attractive to them. Maybe they did not feel safe in them, or comfortable. But I concluded that a certain kind of bird has a particular space in mind, when it goes house-hunting.

The summer before, I had torn apart a wooden deck whose boards needed replacing; when the outside board came loose after prying out its nails, I flipped it over and found half a dozen nests of startled paper wasps. Startled and angry, too. They were out to pay me back for messing with their digs. At first I ran away, letting them calm down a bit. Then I managed to move the board a safe distance from the work site without getting stung. I still had a lot more deck to take apart, though—how long would my luck hold out? I left the job till early the next morning, right at the crack of dawn. That was when I knew that wasps were dopey, huddling for warmth. And I had a can of Raid handy, to obliterate them. Once past that outer board, though—once past the first one—there were no more wasp nests beneath the deck. None whatsoever. Clearly, the underside of decks made for human beings to lounge and sip drinks on was an artificial habitat for nesting wasps. Any board would do, you'd think. But something about wasp instinct told them that the first board—and only the first board—was the place to set up shop. If a person knew that, they would know exactly where to look for wasps beneath a deck.

There was nothing artificial about shagbark hickories, though. They should draw the bats to roost because they were the real deal. Probably they'd *already* drawn some bats, for all I knew. At least a few of them—whether their compatriots were lying dead in caves or not. Right at that moment, for all I knew there could be bats snoozing high above me. And I wouldn't have a clue. Tucked beneath the shingles of bark peeling from these trees. Hanging upside-down, attached by talons that would naturally clench when the bat relaxed. I couldn't see them, but I might as well believe that they were up there catching Z's. Faith being, after all, the evidence of things unseen. It struck me as a serious conundrum, though, doing all this work on behalf of a creature that was black—or chocolate brown—and strictly nocturnal. The chances of actually seeing bats were not that good. But if you couldn't see them, how could you ever know your efforts had succeeded?

Back at the house that night, a neighbor stopped by to visit. Our land is not posted, and he happened to be walking through our woods and saw what I'd been doing: baling twine running at breast height from tree to tree, and all the shagbark hickories marked with bright orange flagging tape. Then he'd seen the golf cart near the edge of the woods, loaded with my logging tools. What gives? he wondered in a friendly sort of way. I explained about the bats, and how I had entered in a contract with the government. How I was going to soon be daylighting hickories so the sun would shine directly on them. I didn't mention my campaign against invasives; I doubted that this sensible person would believe what I'd been doing for the past few weeks. But I did tell him that the nation's generous taxpayers would be paying me to cut several years' worth of firewood. For my own use. That seemed a deal I thought he would appreciate.

"How much firewood you talking about?"

"Nine or ten cords, I think."

He raised his eyebrows. "You can't move that with a golf cart."

I explained I only used the golf cart to commute to work not to actually drive around the forest.

"How you plan on getting that wood out?" he asked me.

I shrugged. "If it's not too far, I plan to carry it."

"Where to?"

"My pickup truck. Parked along the forest's edge."

He made a pained expression. "What you need," he told me, "is to get an ATV."

"Why?"

"Because they get around. They can go most anywhere."

I was aware, of course, that all-terrain vehicles were part of the Vermont scene. Also called "four-wheelers," they seemed to be the ride of choice for all the local yahoos—at least during temperate months of the year, when they couldn't go carousing around on their snowmobiles. Many of my neighbor farmers had an ATV or two, and I'd seen them used to herd stray cattle or ride fences. But I had not thought of them as serious work tools; mostly they seemed to be used by teenage boys for joyriding and popping wheelies. I was predisposed against them—much like I had been predisposed against glyphosate. Now, though, I was being asked to reconsider things. "Can they haul logs?" I asked.

"No, but they can pull a dump cart. Mine can hold a face cord, if I stack it good and tight."

That was an impressive claim; a face cord is one third of a standard cord of firewood. "Hardwood?" I asked him.

"Sure."

"Green?"

"I guess."

"But that's at least a thousand pounds."

"I can pull it, no sweat. And I have a winch on board, in case I get stuck."

"That sounds like a good idea."

"It means I can plow snow, too. The winch picks up the blade and lowers it down." He glanced up and down my driveway, which is long and hard to keep open during winter. "ATVs are great with snow."

"Aren't they way expensive, though?"

"Yes," he said. "I guess they are." He nodded soberly; facts were facts. "But I'd hate to be without one."

I didn't ask exactly *how* expensive, but I figured maybe two or three thousand dollars. I pointed out that I had recently retired from teaching. Money was a little tight. Toys like four-wheelers were just not in my budget.

"Tell you what," he offered. "When you've got some wood you want to move, call me. I'll come down and show you what an ATV can do."

That was an offer that I couldn't well refuse, so I told him I would be in touch before we entered winter. In the days and weeks ahead, I started paying more attention to the ways that ATVs were actually being used on several of the farms around me. They weren't just for tearing up the turf and burning gasoline; some of the time, they had serious work to do. And to build a trail through the woods for an ATV was much less complicated than what it would take to build a road for a truck or tractor. ATVs were only four feet wide, after all. They were low enough to ride underneath branches that emerged from trees at shoulder height. They had sufficient clearance to go over rocks and stumps. They were good in mud, too—even pools of standing water, so long as it was not too deep. And I knew that ATVs had legendary traction.

I didn't let Cheryl know what I was lusting after, once the seed of owning a four-wheeler had been sown in mind. But I started modifying my daily work routine; for the last hour or so of an average day, I'd spend my time carving trails that an ATV could handle. Wide enough, that is to say, for two people to walk abreast—and if Cheryl asked, I'd tell her that was their intended purpose. Hiking trails. Paths to use on cross-country skis or snowshoes. But I had designed them with a somewhat different thought in mind. I wanted to get the woods honeycombed with trails I could use to haul out firewood from anywhere I found it. Sort of like a circulatory system, with artery trails and veins and capillary paths. It might take me several years to fully articulate, but when I was done I'd have a way to get a four-wheeler within twenty yards or so of anywhere in the woods. Or almost anywhere; I didn't need to get into the beaver swamp, and I knew no ATV could scale the back cliff. Otherwise, though, I intended to develop access. And for the time being, it would be my secret.

The trails were surprisingly easy to lay out and clear; I could almost always route them past the older, larger trees without disturbing them. Maybe prune a low-hanging branch or two, but that was it. Mostly I was taking out sugar maple saplings, and often I could do that with a snip of the lopping shears. Sugar maples growing underneath a forest canopy reach a certain modest height—ten to twelve feet, say—and then plateau because they can't make a good enough living to grow bigger. But they can hang on that way for many years. Treading water until some mature tree in the neighborhood finally bites the dust. When at last they get the dose of sunlight they've been waiting for, maple saplings take right off and soon develop into trees. It's a sly, stealthy means of forest competition; it's a game of patience for the progeny of maples. But there were thousands and thousands of these saplings, and to lay out trails meant removing just a few of them. I'd drag the victims to the nearest pile of underbrush from my prior cleanup work, and toss them right on top. Unlike the buckthorn, they could rot down slowly right here in the woods. Where these piles were high and dense, squirrels had already moved in to bury nuts and build themselves protected nests. Habitat, habitat. In the last hour and change of a buckthorn day, I could lay out fifty yards of new forest trail. Do that for a week, and it begins to add up. After a month, I must have blazed close to a mile.

Eventually I'd grubbed all the buckthorn from the ground that I could get out using brute strength and simple hand tools. That meant going back to work with the chainsaw—and, of course, the jar of glyphosate and the paintbrush. This time, though, I didn't start seeing wild colors. Maybe the punch had lost some kick, sitting unattended. Maybe I had just gotten over my initial qualms. Anyway, the stuff did work entirely as advertised. Any stump I painted with the KleenUp soon expired. It looked dead; it *was* dead. Poisoned to the roots. And I figured any stumps I managed to miss would announce themselves the following spring with a growth of shoots. Then I'd cut them off again—closer to the ground, this time—and finish the job. Working with the glyphosate took a lot of concentration, but after a day or two it ceased to be terrifying. It was like the chainsaw, once I'd set aside the manual and started to actually use it in the woods. A lot of things were scary in the abstract, but not in practice.

"Batshit" behavior continued to occupy my thoughts as leaves turned crisp and started blowing off, carpeting the ground. Times when a normally thoughtful person takes an action that is seemingly insane—but does so in a way that is deliberate and conscious. There was an example in a book I wrote when I was thirty, called *The Souls of Lambs*. A farmer is told by his pregnant wife that she's in labor, but they decide that he should spend the day baling hay before taking her to the hospital. When his field work is done, he goes home and finds her dead. That's a dreadful story about misplaced priorities; over time, I'd come to think it dreadful in other ways. Mawkish and overwrought. Embarrassingly sentimental. But I had the story fresh in mind after thirty years because Anaïs—having turned thirty now, herself—had asked if she could use it as the basis for a song. Sure, I told her. Fine with me. It was not a story that was dear to me, anymore, but I was flattered that it somehow struck a nerve with her.

Now the song was finished and was in her concert setlist; "Shepherd" had proved capable of making audiences cry. Styled as a modern-day Child ballad, the song unfolds over a dozen sad verses and relies stylistically on archaic choices in grammar and diction. I had done that, too, in the source text she was riffing on. "Shepherd" was to be the centerpiece track on my daughter's next CD, part of her ongoing effort to stretch the boundaries of contemporary folk music. I couldn't help but take some pride in her accomplishment—but at its core, I felt that "Shepherd" was a fine example of a person who'd gone batshit. What kind of man bales hay while his wife's in labor?

That got me to recollecting something that had happened when Anaïs was born—a couple years after publication of *The Souls of Lambs*. The night before, Cheryl woke at midnight in a pool of liquid. Warm, clear, salty liquid. She was not in labor, but she'd "broken her waters" as the saying aptly goes. The fetus was no longer in a bath of amniotic fluid. So we woke our obstetrician up and asked him what to do. Let's all try to get some sleep, he suggested. If there really were no serious contractions, we could wait till morning before going to the hospital. After several restless hours, dawn broke and I drove Cheryl into town. But rather than immediately check into the birthing unit, Cheryl had her doctor make a quick examination. He proposed that labor almost certainly should be induced; waiting for contractions to begin on their own would risk a uterine infection. Cheryl listened carefully, then thought of several things she had to do *that morning*. There was a grant application that was coming due. She hadn't organized her tax records yet. She was supposed to take part in a conference call. Lying with an IV drip to stimulate contractions would prevent her from getting these important projects done. I was polite, but stunned. My level-headed, sensible wife had just gone batshit. She and her doctor made a deal that allowed her to take the morning off from birthing activities, but by noon she'd come back and agree to be induced.

Cheryl and I went home and took a long walk on the farm, hoping that some exercise would stimulate contractions. No dice. It was near the end of March, almost April; the sun climbed well up in the sky, and the day was bright. I believed in natural childbirth as much as she did, but I also recognized the need to have alternatives. The need to keep an open mind. Every year we wound up inducing a sheep or two who managed to pass a "water bag" of fluid, but then failed to start contracting. Usually the culprit was a malpositioned fetus; normally, the nose and front legs of an unborn lamb act like a wedge to pry open the cervix. If we did nothing in a case of failed labor, sometimes it would kick in by itself and things would turn out fine—but more often, the lamb would be a stillborn. And in the worst case, the ewe herself could die. If on the other hand we stimulated labor with a shot of oxytocin, the ewe might not be thrilled about it but she'd get right to work. The odds of success were in her favor—and in ours. I reminded Cheryl of these facts in a gentle way; by midmorning, she was ready to go back to the hospital and be induced. At the birthing suite, she had the OB nurse install a "high" IV—high up on her arm, that is—in order to keep working on a sweater that she wanted finished before its intended wearer would arrive. After a couple hours on Pitocin, though, she put away her knitting. And an hour after that, we got to meet our daughter.

Well—for Cheryl to say she had a grant application to complete before she could give birth was totally batshit. In its way, as batshit as the shepherd in *The Souls of Lambs.* Then I wondered: were these stories *always* about people having misplaced priorities? Was that the gist of them, the underlying message? Then the mother of all batshit episodes—in my life, anyway—lurched into consciousness. A story so disturbing that I'd put a lot of effort into keeping it repressed. Now it came back to me in perfect detail, and for the first time its meaning seemed clear to me. Seemed unmistakable, for whatever that was worth.

When Cheryl and I had first moved to Vermont, my father spent a few days helping us lay in an underground electric line. We were converting an old hay barn on the premises to serve as our dwelling for the first few years, and though the structure was a stone's throw from our rural road it had never been connected to the power grid. A trench was dug from outside the barn to the nearest utility pole, and a braid of thick wires for underground service was laid from point to point. At the pole, the power company installed a meter box; we had to run each of three fat wires up into the box and clamp them into place. That meant cutting the conductors to exact length, then stripping insulation from each wire's end.

My father marked the first wire at its cutoff point, then told me to pick up a hacksaw and do the deed. He would hold the wire in place between his two hands while I worked the saw back and forth, cutting through it. But I wasn't all that skilled at wielding saws; on the third or fourth stroke, the blade skittered from its kerf and danced along the wire till it found my father's flesh. His index finger, actually. On his left hand. Midway between the first and second knuckles. And I'd cut it to the bone: blood came spurting out, pumping with what looked to be arterial pressure.

I dropped the saw and shouted something like: "Omigod! Sorry, Dad! You stay here—I'll get the car. We're going to the ER."

He gave me a look that was well beyond pained. Also beyond any ordinary rage, and way beyond everyday parental disappointment. "We'll finish the job," he said.

"Fuck, no," I told him. "Look, I cut you with a hacksaw. Your finger's bleeding—bad."

He stooped to pick up the saw with his right hand—the one that was not gushing blood. "Do I have to do it by myself?" he demanded.

Totally batshit. Blood was dripping down his pants, blood was on his socks and shoes. Blood was pooling up on the tufts of grass beneath his feet. I felt awful, but I also knew that he was wrong. Finish the job? Now? Under these circumstances? I'm pretty sure I started cursing him out. I told him that I was grown man, not a child. He didn't have to turn this into one more lesson about the value of work, or the importance of seeing projects through. I'd made a bad mistake, slipping with the hacksaw. *He'd* made a bad mistake by thinking it was smart to use his hands as a human vise. But this was no time for playing mind-over-matter games. He was losing blood, and we were going to the hospital.

He pulled a snot rag from the pocket of his pants—a dirty cotton handkerchief—and wrapped it tight around the wound. "Not until we finish the job," he repeated. In an icy tone, this time. I could see there was no way he'd be backing down. So we spent the next twenty minutes finishing the job: cutting each of three thick wires to the proper length, stripping off insulation, smearing antioxidant goop on the exposed ends, fitting each wire in its clamp and screwing down the nuts. I was shaking, running on anger and adrenaline; I'm not sure which of us was in greater shock. But we got the job done, since that was so important to him. Later, at the ER, they cleaned the wound and closed it up with four neat stitches. They gave him a booster shot in case he'd been exposed to tetanus. And then they released him to get on with the day's work.

Batshit, yes. For sure. And years later, I couldn't help but cringe when my dad would show his scar to people and tell them how he got it. So—was that a story about misplaced priorities? *No*, I now could finally see. He was telling me that when it came to manhood—to a certain kind of manhood, at any rate—I wasn't fit to carry his jock. He must have loved having had the chance to make that point, and make it so dramatically. There was no inherent reason why those wires had to be installed that morning. The power company hadn't even made a date yet to have a lineman come turn on the juice. What if I had cut his finger *off*? I wondered. Would he still have said we had to finish the job?

I had reached the place where when I'd had enough of buckthorn for a day, I'd call it quits. Where the taller, bigger buckthorns had grown up in thickets it was hard to bring them down even using the chainsaw. I could buzz my way through a five-inch stem, and a twenty-foot buckthorn tree wouldn't even shift position—let alone fall for me. All its nearby brethren would pitch in to keep it standing upright, thanks to the way their twisted branches were entwined. Sometimes it was easier to start from the top of a cluster of buckthorn trees, rather than attack the base. That meant working with the saw well above my head—not a recommended practice. Then I'd whale away at random branches till the web of support had been compromised and weakened. That worked—but then those severed branches would collapse on me, spiky thorns and all. My arms were getting cut and bloodied, even through a long-sleeved shirt. My inner mood was often grim, battling the buckthorn. Brendan Weiner had been right: this was a plant from hell.

Eventually, though—around the middle of October—there were no more buckthorns left to deal with in the bat zones. All the smaller stuff had been extracted by the roots, and all the bigger stuff had been reduced to poisoned stumps and enormous piles of brush. I felt whipped—and WHIPped—but I had managed to beat the dread invasive. Compared to it, the barberry was no big deal; nor was the possibly exotic form of honeysuckle. By Halloween, I'd put in a call to George Tucker and invited him to come out and check my work. He proposed a date, and I prepared for it nervously by retracing my steps throughout the bat zones of the forest—almost inch by inch. Cheryl offered me a tip from years of working with government inspectors who would show up at day-care centers, trolling for hazards that could force a center's doors to shut. "They always have to find *some*thing they don't like," she told me. "Otherwise it's like the person hasn't done their job. So even if you've got the woods just the way you want them, you should make sure there's something there for George to find."

"Why?"

"Because he'll need that."

"But I've done a thorough job. The bat zones are weeded."

"Then you ought to maybe change their boundaries by a bit. Move the baling twine to rope in something he can find for you. Then you take care of it, and everybody's happy."

I had to believe that Cheryl knew her stuff when it came to understanding government inspectors. So I took another hank of baling twine to the woods and widened out the boundary of the polygonal bat zone in order to enclose a few honeysuckle plants. There had been so few of these to deal with anyway, it would make sense if I had overlooked a bush or two. Then, too, I recalled that the exotic honeysuckles could only be distinguished from their native cousins by cutting through the central stem to see if it was hollow. If George Tucker did come across the honeysuckles, there was still an even chance they'd turn out to be native. Then he could conclude that I'd become an ace at botany.

Came the day, and I put George on the golf cart and drove him to the woods. He had a spiffy new GPS device with a dish-shaped antenna poking up from a knapsack and rising well above his head. It kept hanging up on overhead branches, forcing him to duck or stumble every so often—but he said its readings would be accurate to within just a couple feet. Almost as good a device, he told me, as the military gets to use. The long runs of baling twine I'd used to mark the bat zones impressed him; he had never seen that done before. Then, too, he liked my use of flagging tape to show off each and every shagbark hickory. When it came to measuring the long, narrow polygon, he programmed the GPS to monitor his footsteps as we walked the full perimeter. Then it would report back on the area enclosed. It took a good forty minutes to follow the windings of the baling twine from tree to tree; at one point George lost his footing on a stretch of steep ledge and nearly broke his GPS. But it kept recording data, and eventually it told us that the long, narrow bat zone was two and four-tenths acres. Adding in the five half-acre zones as well, the grand total was just shy of five acres.

"And you'll pay me for that extra acre?" I asked hopefully. "The contract only called for four."

"Sure, we'll pay," he said. "You've done a really good job here. I wasn't sure what to expect, but you're on top of things. This is what we like to see."

"Thanks." I felt my chest swell—and on account of what? On account of two months of crawling through the woods like a snail. Or a slug. "It wasn't easy," I said.

"Took a lot of time, did it?"

"I was keeping track at first, but then I said to heck with it." Now I made a guesstimate. "Three hundred hours, at least. Maybe more like four."

"Well, I have some good news for you. Next year will be easier."

Next year. Right—I wasn't done with this yet. I had another year to go before the forest could be opened in ways that would let sunlight warm the shagbark hickories. That's when the bats would supposedly start moving in. "How are the bats doing?" I asked. "Are there any left?"

"They're having quite a time of it, from what I hear. Not just in New England, either. White-nose has been showing up as far away as Tennessee. I even heard—wait. What's that over there?"

"What's what?" I asked as innocently as I could. We were approaching where I'd recently reset the twine; George had found one of the honeysuckle bushes that I had deliberately enclosed. "Damn." I shook my head. "My bad."

"You know what this is, right?"

"Yes. And I promise I'll take care of it. Today."

"First we ought to check something." George grasped the bush's central branch and snapped it open; inside, the honeysuckle's stem was firm and solid. "This is native," he said. "So you'll want to let it stay."

"Will do." I felt like opening a beer with him—or possibly a six-pack. But it would have breached my promise to maintain a drug-free workplace. "So—you think I pass?"

"Yes, you pass—and then some. You should get your money in the next two weeks. If it doesn't show up on your bank statement, let me know. I'll take care of it."

Now it seemed possible to talk out of school with him, and I had a couple of questions that I'd been postponing till I'd gotten my report card. "What about the Norway maples?" I asked gently, so as not to raise alarm bells. "Aren't they considered an invasive species, too?"

"You don't have Norway maples here," George assured me.

"Brendan Weiner said I did."

"Did he?"

"He said they're hard to tell from sugar maples, but I have some. And that if you made me cut them down, I'd have had my hands full."

"I don't think Brendan Weiner knows what Norway maples look like. Anyway, as far as we're concerned they're not a problem here. You've done what we asked you to—leave it at that."

But I wasn't finished yet. "Do you think we'll win against invasives, in the long run?"

"Probably not."

"Me neither. I mean, look at what I did—how many landowners are going to do that? And even in these woods, it's just a drop in the bucket. Five acres out of what? Seventy, at least. And even on those five acres—will a bit of garlic mustard matter to the bats? I don't see it. I don't think so."

"All we can prescribe is what our guidelines say we have to. Invasives need to be controlled before you open up the woods. In situations where that didn't happen, things got worse."

"Well," I said, "it's been some experience—that's for sure. One I never thought I'd have."

"I'll bet you know these woods, now."

"Yes. I know them intimately."

We *did* leave things at that, and my next bank statement showed a direct deposit of $1,127 from the U.S. Department of Agriculture. Four dollars an hour, at best, to rid five acres of buckthorn and garlic mustard. But it came to something—and at least I'd had a project to obsess about for all those weeks. When I closed my eyes, I could go for walks around the woods. That's how well I'd come to know them, and the knowledge satisfied. Then, too, I was starting to address some issues that had long been gnawing at me. *That* seemed promising, although I had a ways to go. And I was becoming a different sort of person, too: almost post-literate, and sworn to taking hands-on action rather than considering ideas in the abstract. Maybe doing all this work for WHIP was truly batshit, but each time I looked into the woods I had to smile. That was an unexpected benefit to pulling weeds. That was so much gravy.

Niche

The winter of 2010–11 arrived with an extended spell of mild weather. Early in the week of Thanksgiving—when there *should* be snow—my neighbor rode down through the woods on his ATV. I was burning piles of buckthorn, trying to keep three or four fires going at once; a haze of smoke lingered in the air, but the ground was damp. I had no fear that things would get out of control. In the woods were several cords of firewood stacked where I had cut them in the past few weeks. Steven—who had helped me make hay in his teenage years and now was a father of teenagers, himself—told me to hop onto the saddle behind him. This is not a recommended practice, but neither was the fact that we weren't wearing helmets. Or thick gloves, or boots. There were decals all over the side of the ATV with warnings about what could happen. What was safe and what was not. Anyway, I pointed out the entrance to my main trail through the nearby woods and we cruised it at speeds that made my hair stand on end. Rough terrain was not a problem. Anywhere I'd stashed a pile of wood was within a few steps of where the ATV could go. When Steven pushed a switch that engaged the four-wheel drive, the Suzuki turned into a tank. We went riding over boulders and tree stumps—even deadfalls up to several inches in diameter.

"Wow," I said. "That's something."

"It helps to have the oversize tires," he told me.

"How big is the engine?"

"Six hundred—but I think that five would be enough."

I knew he was talking about cc's of displacement. When I owned a Porsche 912, the engine's size was 1.6 liters. Sixteen hundred cc's. That was a car, though, and it was possible to drive it at alarming speeds. Six hundred cc's in an ATV seemed overkill. "I'm not out to break any speed records," I told him.

"But you'll need the horsepower, if you're hauling firewood."

That was a consideration; we were just joyriding, not doing real work. Anyway, I was thrilled to see that my forest trails could be negotiated by a vehicle like this. Now the issue was how I was going to get my hands on one. Meantime, I was stuck using my back to carry wood from where I'd cut it to as close as I could drive the pickup truck. I'd learned to haul out heavy sections six to eight feet long, then work them up into stove length at the house; that was much easier than stumbling through the woods with an armful of shorter pieces and, more times than not, dropping some along the way. Still, it was a bullheaded way to go about things. I was no spring chicken, and I knew that eventually I'd have to act my age. Either that, or wind up wearing a support brace.

By mid-December, I had the requisite three cords of firewood cut and split and neatly stacked inside the garage—and at least another three cords were piled in the woods, waiting till I had the time and energy to get them out. And much of this wood had come from deadfalls lying in the bat zones, many of them dating to the Ice Storm of '98. Looked at in a certain way, I was a rich man. In a state where keeping warm in winter can be quite expensive, I seemed to have a lock on cheap and abundant fuel. And, by cleaning up the woods, I was helping bats. My forest trails would serve as flyways guiding them to roost trees. Once they found them, they could crawl beneath the bark and move right in. They could raise their pups in a Hilton, rather than some low-rent welfare hotel. And the bat zones on our farm were easy on the eye again, pleasant to go strolling through and fun to show to visitors. All things considered, the results of my efforts seemed a win-win-win.

Did I say the weather had been mild, as we entered winter? Prior to the middle of December, there was little snow or cold to deal with. But that didn't last; by Christmas, all hell had started breaking loose. By the season's end, over thirteen feet of snow had fallen—nearly twice as much as in a "normal" year, and enough to make it qualify as the third snowiest winter in Vermont. Or in the history of Vermont weather records, which go back to well into the 1800s. Keeping our thousand-foot-long driveway passable became a real challenge. And we *had* to keep it open, since visitors were sometimes coming to The Annex and we could not guarantee that they'd arrive in cars equipped with four-wheel drive. Or even snow tires. Then, too, Cheryl's parents had become advanced in age and ever more frail; there was a chance we'd have to call for an ambulance some bitter, stormy night and somehow get it to their door. That would be impossible if snow had closed the driveway.

Dealing with six-to-eight-inch snowfalls became routine, handled primarily with a John Deere scraper blade mounted on the back of our utility tractor. Plowing snow from a tractor's rear is always problematic, since much of the white stuff gets flattened by the tires' lugs before the blade can pick it up and shove it to one side. But most tractors come equipped to handle basic implements mounted at the rear, and having these mounts—which include an internal pump to generate hydraulic lift—on a tractor's front end costs a lot more money. The rear blade works well enough under most conditions, but a really major storm can render it useless. We had major snowstorms nearly every week, that winter. Storms of fifteen inches, twenty inches, and more. The tractor's belly rides fourteen inches off the ground; when snow piles up much deeper than that, all the machine can do is wallow in a sea of fluff. Rather than plowing through the drifts, it sits there treading water.

Plan B involves a pair of heavy-duty snow blowers, but they only clear around a two-foot swath per pass. After some of the heavier storms, Cheryl and I would run them both for several hours—back and forth from house to road, gradually gaining the width that would be needed to get vehicles in and out. When the snow was wet and heavy, working with the snow blowers was a slow, painful grind; when the snow was light and fluffy, winds could drift the snow back into the driveway with surprising ease—even long after the storm had passed and skies had cleared. Trying to blow snow under gusty conditions meant constantly adjusting the direction of the discharge chute, and you wound up looking like a Yeti anyway. It was hard to dress in sufficient layers to beat the cold; some days, the temperature would top out at five below. Just to get the tractor up and running under those conditions meant warming the cylinder head with a hair dryer and using a charger to goose up the battery.

Dealing with the driveway again and again made me come to appreciate the way bats deal with winter. Hibernation struck me as a marvelous adaptation. Caves are never warm, but on the other hand they don't freeze up. Once perched upside down with thousands of its kin, a bat's metabolism dials down to almost nothing. Body temperature cools to nearly that of the surrounding air; heartbeat and respiration slow until the animal enters a state of torpor. Long winter's nap, indeed. Then when spring arrives, something kicks in to stir the bats back to life again. Maybe they feel groggy at first—and really hungry, too. But they have managed to skip out on months of weather to which they are not adapted. Weather that would quickly kill them, if they were exposed to it. Some bats are known to live for thirty years or more—half of which time they spend snoozing in a cave. Thirty years of summer, then. Thirty years of never plowing snow, or keeping fires stoked, or otherwise having to do battle with the elements. I'd say they are onto something.

My preferred adaptation to winter is to curl up close to the woodstove with a book. And since my life was fast becoming all about bats, I started seeking out whatever I could read about them. When our kids were young—before the advent of the Internet—we got them an old-school encyclopedia, which they turned to many times per week. Now it sits languishing—all twenty-two volumes of it—on a dusty shelf. But I pulled out volume **B** on a cold winter day and looked up what *The World Book* had said about bats in 1989. For starters, I read that there were more than nine hundred different species of Chiroptera. That seemed an awful lot. And volume **M** gave the total number of mammalian species as being four thousand, give or take a few. But that meant that twenty-two percent of mammals were actually bats—an incredible percentage, beaten out only by the members of Rodentia. Between my "flying rats" and the real ones—the rats that could not fly—these two orders had the class of mammals sewn up. But what could it mean that one-fifth of mammal species were some kind of bat? Imagine that much evolutionary energy expended on an animal whose hands had first grown huge and then acquired webbing stretched between the fingers. Some tough, ultrathin and flexible membrane. Presto: now there was a mammal that could fly.

Donald Stokes was a classmate of mine in college; I've long admired the series of nature guides that he and his wife, Lillian, have authored and edited. So I sought out the Stokes's *Beginner's Guide to Bats*, published in 2002. That book gave the number of bat species as being "nearly 1000"—a significant bump from what *The World Book* had pegged it at, thirteen years earlier. If that was an accurate figure, then it meant that in the brief time between 1989 and 2002 one hundred new species of bats had been discovered. And then I came across Phil Richardson's volume *Bats*, published by London's Natural History Museum in 2011. This gave the number of Chiroptera species as "more than 1100"—which, if true, meant that another hundred species of bats had come to light. Then a quick check of Wikipedia's page on bats put the number at 1240 species, amounting to another jump of more than ten percent. You'd think that bats were doing pretty well, considering. We often speak of species extinction as a modern ecological crisis and a bellwether of our overall predicament; in environmental circles, it is commonplace to hear the loss of biodiversity decried. But here was a cohort of mammalian life whose numbers—or at least whose number of known species—seemed to be expanding relentlessly. And this despite the widespread fear that bats were now endangered. What on earth was going on?

Digging deeper, I learned that to some extent the shifting tally had to do with hair-splitting among finicky taxonomists. Someone finds a reason to make two species out of one, based on some previously unobserved feature. Since bats are damn hard to observe in the first place, it wasn't shocking to learn that a few details might have gotten overlooked. The rake of a calcar, say—the tiny spur of cartilage poking out between the hind legs and the tail of a bat. The rake of a calcar could result in determining a hitherto unrecognized species. It was a game that chiropterists liked to play. Then, too, modern genetic science has had an impact on the practice of taxonomy. With DNA profiling and whole-genome sequencing, species that formerly could not have been distinguished from some other species are teased out. But even so, the number of bat species was astounding; by contrast, the Primate order—despite its clear accomplishments and seemingly high rank on the zoological totem pole—is comprised of only about two hundred fifty species. Primates are far less diversified than bats.

When I raised these issues with a former teaching colleague—an expert on matters of environmental science—he nodded at the bats' success. "They must have had one hell of a niche," he said. And he didn't mean a recess built into a wall; he meant an *ecological* niche, which is based on understanding how a given organism makes its living in the world. How it takes advantage of certain opportunities and manages to dodge a variety of bullets. For example, if you're a large browsing animal and the food is overhead, you'd be smart to stretch your neck and make like a giraffe. Do that for long enough, and the species that we call *giraffe* will have evolved. If you are a monkey and you live on fruit that grows in trees, having a prehensile tail is a great advantage. After several million years of monkey evolution, there won't be any monkeys making do without them.

For most bat species, insects represent the primary source of food. There must have been a lot of insects to feed upon, back when the bats were first evolving. Nocturnal insects, too. Enough of a food supply to encourage adaptations that resulted in a mammal that could catch and gobble bugs in flight. Hunters who could harvest dinner out of thin air—and do so with great success, time and time again. Flying by night was a hedge against the odds of being taken by a predator, just as hibernation was a hedge against the cold.

Bats are presumed to have arrived on the scene between fifty million years ago and thirty million years ago. By the latter date, based on

DNA evidence all twenty-six known families of Chiroptera had come into existence. That's a long time before even the great apes, let alone Neanderthal versions of ourselves. One thing struck me as pretty well certain, though: had the U.S. Fish and Wildlife Service—which is empowered to make these decisions—been in business when the bats arrived, surely they would have been condemned as an invasive species. All of a sudden, flying insects were not safe at night. Call the cops—the ecological equation had been altered. But that only brought to light the fallacy involved in making these determinations; it's as if the experts had taken a photograph of the environmental state of affairs at a particular point in time, then declared that deviations from that norm should be resisted. Or reversed, if possible. But natural history is not a static proposition, and it long precedes our own arrival on the scene. Humans have been around to regulate the ecosystem—or at least to try to—for only the briefest time. In some ways, the whole show was doing pretty well without us. Maybe it was doing fine. Conceptually, I did feel the bats deserved a helping hand from humans in the face of this white-nose syndrome; that was a very recent crisis they were challenged with, and their ability to fight back was limited. Spurred on by the government, people like me were justified—perhaps—in taking measures that might help them out. But if bats had only just arrived at the party, I'll bet we'd be working just as hard to get rid of them.

Bats had a remarkable adaption to the needs of reproduction, I learned in my studies. Female bats breed with males at the height of autumn, before going underground to spend the winter months. But a cave bat's typical gestation is only six to eight weeks long, and there's no way a bat could give birth successfully in the midst of hibernation. For that matter, what pup would want to have a mother who was constantly nodding off? And then there's the issue of not having any food to eat. So female bats set the male's seed aside and store it up internally—somewhere near the uterus—until summer rolls around. Then, once they're back in leafed-out forests and awash in bugs, the female finds that stored-up seed and uses it to fertilize herself. When the time is right. In other words—to put things bluntly—lady bats get down with the males when they're in the mood, but they don't get pregnant till they're good and ready to. How many women wouldn't sign up for a deal like that?

What kept coming through again and again, though—as I read up on the remarkable ways of bats—was that these little guys were mammals like ourselves. And as fellow members of a shared class of organisms, it seemed to me that perhaps we all shared a niche. What are the advantages that mammals are seizing on, doing business as we do? Well, we all have hair on our bodies—for what that is worth. Hair helps to keep us warm; it gives us something we can groom—or get someone else to groom, if that's what we are into. We all have some tiny bones—auditory ossicles—hidden in the middle ear, that amplify sound in a way that dramatically expands our knowledge of what's going on around us. We all are born alive, in a shared performance that requires both mother and fetus to cooperate—to work together toward the common goal of parturition. And we all drink milk as babies, suckling from our mothers' breasts. That in itself is a huge bond connecting mammals. Huge enough that humans ought to see the bats as relatives.

The genius of milk is that it offers a complete solution to the challenges of infant nutrition. It's a high-octane fuel, easily digested and nutritionally tailored to the needs of a given species. And it's available virtually on demand, presented at a constant temperature and constitution. Day and night. Round the clock. Whenever the baby's hungry—but milk quenches thirst as well as satisfying hunger. Then, too, milk affords a quick inoculation against pathogens specific to the animal's environment. If a mother has created antibodies to disease, those are passed on to her baby in her milk supply. And for at least the beginning of a mammal's life, its body is capable of harvesting antibodies directly from the gut. That capability shuts down before too long, but while it lasts it gives the young a leg up on infection.

Perhaps most important, though, is that lactation offers the profoundest sort of context for nurturing behavior. Mammals have a way of taking care of each other—more so, I would argue, than snakes or sharks or spiders. More so than a tree, for sure—not to mention garlic mustard. For a milking female, suckling is also a symbiotic interaction: she *needs* to have her mammaries drained of milk, just as her baby needs the goodness that it brings. Otherwise mastitis would inevitably set in.

Male bats are not involved in nurturing their offspring. For that matter, male bats are probably clueless as to who their offspring are. But this is not unusual in mammal species. The two rams we keep on hand to breed our flock of ewes show no interest in their yearly crop of progeny, let alone a sense of being partners in raising them. A son—if it *is* his son—becomes of interest to a ram only when it's big enough to take a chance at butting heads. A daughter—if it *is* his daughter—becomes of interest to him only when she comes in heat and stands for him to mount her. But male sheep do at least live in a community where child development and nurturing are going on. They all graze together, drink at the same trough together, lie down and rest together. Rams do exist within the context of a flock. Bats, though, segregate the sexes at birthing time. Females gather in maternity colonies, whereas the males tend to go off by themselves. If a shagbark hickory tree gets chosen as the home for a maternity colony, hundreds of bats will make their roosts there till the pups are grown. Male bats are not allowed. If there's rhyme or reason to this matriarchal strategy, something about male bats must make them unreliable around their offspring. When it comes to reproduction, they are good for just one thing.

My father was actually a nurturing sort of male when I was a youngster, despite the many challenges I now know that he faced. I remember being carried from the car with tenderness, half-asleep after one road trip or another, and tucked gently into what he'd call *your own little bed*. I remember being read to, zippered up in warm pajamas. I remember watching him make animal pancakes, pouring batter into shapes that cooked up into chubby bears and long-eared rabbits. I remember cocoa and S'mores around a campfire. Thinking of my own life, I would hate to have missed out on helping Cheryl rear our kids. Sometimes, to be honest, it was tiring and tedious; there were moments when it seemed an utter waste of time. But mostly I enjoyed hanging out with Ethan and Anaïs, sharing their imaginative worlds and finding ways to let them enter mine. I would have hated to be banished from some human version of the bats' maternity colony, forbidden to approach the place where children were being reared. But then human culture is defined by our extremely long childhood; it takes two to raise a child—or the work goes better that way. That's what you get with a mammal whose brain at birth is less than half of its eventual size. Were it any bigger at birth, though, babies would not be able to get out. They have a hard enough time getting out already. When it comes to pinning down the niche that humans occupy, it begins with seeing what a giant brain can do.

As the winter wore on, the concept of niches proved a fertile line of meditation. I would return to it again and again. *People* had niches, too—I mean as individuals. For example, I had a niche working in the woods. I had certain tools that conferred adaptive benefits, increasing my chances of success at getting firewood. I had a darn good chainsaw, and a pickup truck. The opportunity was there to heat my house indefinitely—and at no significant cost, compared to fuel oil. What were the bogeys I was trying to avoid? Well, the work was dangerous and I could not make it safe. It would be easy to hurt myself seriously, either from an accident or repeated insults to my aging back. The latter seemed entirely likely, given the effort involved in carrying wood from the inner forest to where I could park the truck. I was not adapted to perform this sort of labor; it was an ongoing problem with my niche. But I had quiet hopes of changing all that. Snowshoeing with Cheryl on a sunny winter afternoon, following the woodland trails I'd blazed the fall before, I dropped the hint that I had built them wide enough to accommodate an ATV.

Cheryl didn't go ballistic. "Now we're getting one of those?" she asked me in a neutral way.

"I thought I'd shop around," I said. "See what's available."

"What do those things cost?"

"I have no idea, really. Three or four thousand, new? But I'll try to find one used."

"And you really need this?"

"I have seen what they can do. We've got tons of firewood lying on the forest floor, but it's starting to go punky. It won't keep outdoors forever. With a four-wheeler, I could drive to where the trees are down and load up a dump cart. Then haul the wood out and store it under cover."

Cheryl listened patiently. On the one hand, I had stopped bringing home the bacon in our domestic economy. *Any* of the bacon—but that was a choice we'd made. We were far from destitute, but there were financial pressures and it would be easy to regard an ATV as an unnecessary luxury. Even as an affectation, as if I were trying to prove my masculinity. I could imagine all these thoughts running through her head. On the other hand, my wife was well aware that I'd become enthusiastic about working in the woods. True, some days were boring and by evening I would be worn out. But I had been happier, the previous fall, than I think Cheryl had seen me in a long, long time. If a four-wheeler had become the tool of my dreams—and if it *was* a tool, able to advance my scheme of cleaning up the forest and harvesting firewood—then she would be foolish to prevent its acquisition. "We don't have the money," she said. "But see what you can get."

Suzuki was the brand of ATV that my neighbor drove, and that he had recommended. They had a dealership just outside of Middlebury. It was primarily a motorcycle showroom; when I walked in, head-banger music was blaring at high decibels. Chromium and black leather was the ambient décor. "Welcome to the toy store!" said a friendly man with sideburns. "What can we do for you?"

I tried to put a game face on, but I could sense it wasn't working. I hadn't been on a motorcycle since the sixties, and I didn't look like I was going to ride one now. I said I'd come to check out ATVs in case he had some. I told him that my neighbor liked Suzukis, and had told me they were great for working in the woods. He'd said I should buy one with at least five hundred cc's—an engine big and strong enough to pull a good-sized cart, and maybe plow snow. I said buying used was fine, in case they had a trade-in.

"Only new," he told me. "People ride these till they fall apart."

"Can I look at what you've got?"

He walked me past the rows of sleek, brightly colored motorcycles—bikes that looked like they were breaking speed records standing still. Then he brought me face to face with a canary-yellow King Quad. The color of my long-lost Porsche. It had power steering and an engine of the right displacement. It had macho tires and a rugged-looking, brute physique. And it had a sticker price of eight thousand dollars. "Ouch," I said. "That's a lot."

"Actually, it's on sale. Only this month, though."

That kind of money was out of the question, but I kept the conversation going out of curiosity. Matching the King Quad to a cart for hauling firewood would run me seven hundred more. Mounting a snow plow engineered for the machine would cost another thousand—but then I would need a winch to raise and lower the blade. Plain vanilla winches were around four hundred dollars, plus installation. Getting the right sort of helmet was important, too; that would run at least a hundred. Then there'd be a six percent sales tax and a nineteen-dollar fee to Vermont's Department of Motor Vehicles for a certificate of title.

"They title ATVs?" I asked.

"Sure. They title snowmobiles—motorboats, too. ATVs are vehicles."

"But they can't go on the road."

"Doesn't matter."

Adding it all up, I'd be in for darn near eleven thousand dollars—considerably more than I had paid for a yellow Porsche back in 1970. And the ATV could only seat one person, legally. It *did* have a speedometer that went up to seventy miles an hour, but I didn't think I'd ever push it over twenty. The salesman punched the numbers into a computer and handed me a printout of the deal he was offering. Then he asked me how much I was planning to finance. He could fix me up with an attractive rate of interest, if I would agree to buy insurance from him, too.

I raised my hands in a *back-off* gesture. "I'm shopping around," I said. "Just getting a feel for things."

"Would you like a test-drive?"

"Maybe some other day."

So I booked it out of there and drove home, my wings clipped. Who could afford to drop eleven thousand dollars on a tool for hauling firewood? I was too embarrassed to tell Cheryl where I'd been, let alone what I'd learned about the cost of ATVs. But then, poking around on the Internet, I found that Suzuki's prices weren't far out of line; Honda, Kawasaki, and Yamaha were all selling similar machines for about the same money. Then it struck me that these brand names all had one thing in common. Had everybody in Japan taken up four-wheeling? There was one American brand that came up on my screen: Polaris, based in Minnesota. I knew that they sold a well-regarded line of snowmobiles, but it turned out they were making ATVs, too. Not that their price point was notably better, but the Polaris machines were somewhat heavier and came with bigger engines. Some of them even had two cylinders, instead of one. Able to burn more gas, as we like to do. But I had to shelve my dream of owning any such machine—not unless one fell off a truck and landed in my lap. My niche in the woods was not likely to be enhanced; I would be stuck with a perpetual backache.

Human personality was open to analysis in terms of niches, too. It involved the same considerations as applied to pinning down the way a given species occupies its space. What opportunities are being sought by everyone? Not, in our advanced culture, adequate supplies of food. Not basic shelter, either—not, at least, for most of us. People seem to hunt for love, primarily. From other people, more times than not. And a sense of personal empowerment, too. What are the fundamental risks each person faces? Separation from affection—in a word, loneliness—and the fear of looking weak or foolish in another's eyes. Humans have myriad ways of winning love and warding off rejection, tailored to each person's own psyche and situation. Hence the rich drama of human personality. The strategies that work for children when they seek to win the affection of others—parents, siblings, friends—are bound to be repeated as they move into adulthood. Ditto for the strategies that make them feel competent, able to make things happen in the world. Ways of behavior that do not achieve these goals will be discarded, and perhaps condemned. But since each person's experience of life is essentially unique—different, in important ways, from that of any other—each of us carves out a somewhat different niche. Once we do carve it, though, we occupy it faithfully. Each person clinging to behaviors that have worked for them.

When my father died, my son—Ethan—found an autobiographical document tucked onto a shelf above his personal computer. Then my sister found the file saved on his hard drive. "To the Grandchildren and their Parents," it began. "It occurs to me that at some time you might want to know what your origins were." Some of what followed was read aloud among the family members after my dad's funeral, but at the time I felt put off by what struck me as a pack of lies and half-truths. Now, though—sometime in the winter of 2011—I dug those pages out and read them over carefully, trying to gain insight on the niche my father occupied. Clearly he was trying to reveal things he'd never shared; equally clearly, though, he wasn't coming clean with us. You had to read between the lines to see what he was saying—or was trying to say, but couldn't quite admit. But reading in between the lines wasn't all that hard to do, once I got the hang of it. It was like a palimpsest; the underlying facts showed through.

For example, by my dad's account, when his parents got married his father, Guy Treleven, was thirty-seven years of age and his mother was nineteen. Talk about cradle-robbing—but it was a different era. The wedding took place on Christmas Day of 1913, in rural Wisconsin. Wait, I thought. Christmas Day? Who gets married on Christmas Day, and why? The bride had been working as a housekeeper on the Trelevens' multi-generational farm; the Trelevens had come to Wisconsin from Cornwall sometime in the 1830s. Guy told people he had never married before because no woman had ever appealed to him. My father wasn't born until nineteen months later, but another child before him had been born and then expired. Reading this, I smacked my forehead. Two full-term pregnancies in the space of nineteen months? Not likely. Not even possible, for most women. Can there have been any real doubt the bride was pregnant?

Lillian Hollander, my father's mother, had recurring bouts of what today would surely be called clinical depression. Getting impregnated by a man twice her age and then losing the baby couldn't have helped, and yet by my father's account—but how could he have known?—she had been mentally unstable before that. After my father's younger brother, Earl, had been born, Lillian entered a prolonged melancholic spell. During lifelong episodes when she simply couldn't cope, my dad's mom was sent away to live with *her* mother. Della Hollander comes across as a stern and somewhat cold woman, someone who would not be sympathetic to what she would have deemed a mental illness. My father wrote of her that if his grandmother learned somebody loved a certain kind of food, she'd get them to eat enough of it to make them sick. No doubt it was fortunate that Lillian had someone she could turn to when she couldn't function, but Della's care for her—callous and guarded as my father hints it must have been—seems unlikely to have been therapeutic.

A fourth child was born but died, bringing on another case of postpartum blues. By now Guy Treleven's farm had failed and he'd moved the family into town; for a while he managed a creamery there, but the board of directors turned against him and forced him out. Soon there was no money, literally. None at all. Eight years into the marriage, Guy's health failed; when my dad was nine years old, his father died of cancer. Fifteen months later, Lillian remarried to a man whose job was teaching farmers how to keep financial records and calculate depreciation. Donald Mitchell worked for the state university in Madison, and he was willing to accept a widow saddled with two young boys. But he insisted that they change their surnames to his own, and they were prohibited from mentioning their father.

My Gramps's taking on the care of this unlucky family rarely was discussed with us, but when it came up it was held to be an act of supreme generosity. Reading in between the lines of my father's text, though, a different rationale emerged. Gramps was a "clean" man; he didn't drink or smoke or swear. And he seemed to think the act of coitus was unclean as well. Lillian was clearly capable of impregnation, but my father recollects that three years into the marriage his stepfather was teased for not having sired a child of his own. Perhaps in response to this, he did have a son by Lillian in 1930—after which his wife had a full-on nervous breakdown. Without getting too far ahead of myself, I need to say at this point that I think my Gramps was far from straight in terms of his sexual orientation. Not that *straight* would have been a word in his vocabulary, nor would he ever have used a word like *gay*. But something about the thoughts and sensations that aroused him was askew. And had this been known about him, in that time and in that place, it might have been quite damaging to Gramps's career. Marrying a widow with two children was perhaps a way to rise above suspicion and attain respectability.

My father also wrote about his lifelong battle with a speech disorder—something that no one in the family was allowed to mention, let alone ever call direct attention to. Speaking in conversational tones, my dad would often stutter; sometimes his voice would "block" for several awkward moments. The cause of this had always struck me as being more psycho-emotional than physiological, and now—reading this document—my hunch was confirmed. He had been raised in a way that must have sown mistrust as to whether the grown-ups in his life would behave reliably. And whether they would have his back. Like any child who stammers, my father was ridiculed mercilessly by his peers. But after his father's death the family moved to Fond du Lac—a new and much bigger town—and in his new school my dad was cast to play The King in some fourth-grade drama. He was coached to speak in a loud, authoritative voice—and he found that when he did, he almost never stuttered. Hence the adoption of what became my father's preferred adult persona: speaking at people in a loud, authoritative voice that managed to incorporate a note of intimidation. It was a significant dimension of his niche. It made him feel powerful—but equally important, it served as a hedge against his fear of looking weak and foolish. But it made it hard to have a normal conversation with him; he was on safer ground telling people what to do—or warning them what not to do. His affect was a mask designed to keep himself from stuttering.

Lots of people who are drawn to telling others what to do wind up working for the government, I realized. That's their natural arena—their niche— and it helps explain why government can't stop impinging on the lives of ordinary people. Do this, do that. Do not do some other thing. I am not a libertarian, by any means. But I can't believe it was within the founding fathers' vision for our great republic to have a bureaucracy of wildlife biologists telling me I'd have to put in weeks of pulling garlic mustard before I could daylight shagbark hickory trees to help the bats. Some kind of line had been crossed; something was out of whack.

My father should have had a job with the government—he was well adapted for it, and he would have fit right in. I recalled accompanying Cheryl to Washington on one of her many trips to work on child-welfare issues. On the subway, on the street, in the lobbies of hotels was an army of well-dressed men and women, each of them bent on convincing the government to make people do something that they didn't want to do. They were in their element, and I was clearly not. It triggered a flare-up of my Vietnam War resentment from forty years before—and now a certain memory floated up into consciousness and made me want to join a protest, long after the war had ended. March on Washington again.

My sophomore year of college, in the fall of 1966, the Pentagon decided that some draft-able young men might be hanging out in college for the student deferment—even though they lacked real competence as students. So a company was hired to draw up a standardized test for college men to take; if a testee couldn't make the 50th percentile on this test and if his grade point average was lower than the median, the government could cancel that lad's 2-S deferment and ship him off to boot camp. And if a young man refused to take the special test, the same fate would befall him. No one at my top-notch school was apt to fail this exam, but the demographic section led to howls of laughter. Given a list of fifty fields we could check as being our college major, the first box read "Humanities." The second box was "Liberal Arts"—as if there were no overlap between the two, and as if at schools like Swarthmore most students wouldn't have a major that would fall into one of those two categories. The third suggested major was "Aeronautical Engineering," and the next forty-seven choices were similarly focused on the military's needs. "Fluid Mechanics." "Jet Propulsion." "Atomic Science." Without directly saying so, this test sent the message that we ought to be majoring in physics, not philosophy. Chemistry, not classics. If we understood what was best for the nation—and indeed for ourselves.

At the time, it struck me as outrageous that the government would try to nudge me toward a certain field of academic study. As if since they couldn't draft me, they still wanted to control what I could do. Now, though—in hindsight—it was simply one more case of government doing what governments always do, given that they tend to be magnets for people who like telling people what to do. Or, even more rewarding, telling people what they *can't* do. The episode afforded solid training for coming interactions with income-tax auditors and zoning administrators; it helped train me to comply with arcane building codes and the demands of motor-vehicle divisions. It was preparation, too, for dealing with county foresters and LIP directors and agents of a program known as WHIP, under the aegis of the NRCS, within the hydra-headed USDA bureaucracy. There was no conspiracy afoot here, really. Government occupied a certain kind of niche, and thus it couldn't help attracting employees who shared its values. One of those values was a lusting for autocracy—or at least the exercise of autocratic will.

It made me feel rebellious, though. I thought of Marlon Brando in that film *The Wild One,* suited up in his black motorcycle jacket. "What are you rebelling against?" a pretty woman asks, and he answers "Whatta you got?" with an attitude. I owned a black leather jacket myself, picked up by my son when he was living in Baltimore and rehabbing row houses. What I didn't have was a badass motorcycle—which got me to thinking about ATVs again. I had sold some hay, that fall, by posting a classified ad on Craigslist; now I searched their site and found four-wheelers were listed amid camping trailers, Winnebagos, and other "recreational vehicles." One of them seemed to be an outright steal: a Polaris 550 with a snowplow and a winch, offered at the fire-sale price of $2,700. "The best four-wheel of them all," read the poorly drafted copy. "And this is made right here in America." There were some accompanying photographs, and in them the machine looked mighty sharp. The seller was in Waterbury, fifty miles away. But when I e-mailed to express my interest, I entered what quickly turned into a swamp. I proposed driving up to check out the merchandise—kick the tires or whatever—before making up my mind, but I was told such an inspection was impossible. The seller was a woman who'd been recently divorced and had received the ATV as part of the settlement, but she had to sell it fast to pay off her attorney. So the deal was this: I was to place the full purchase price on deposit in her eBay account, after which she'd have the machine shipped to me for a four-day trial period. If I wasn't satisfied, I'd pay to ship it back and my funds would be returned.

I thought about that, and I thought it sounded fishy. So I wrote back and said it surely must be possible to look at the machine at wherever it was being stored. *No,* came the quick response. There was no way such an inspection would be possible. After her marriage's recent tragic ending, the woman had moved out of state. The ATV was crated at a shipper's locked warehouse, and thus could not be seen. But she was a perfectly trustworthy person; all I had to do was ask eBay about that. Look, I said, I'm sure that you're a trustworthy person but I don't feel comfortable doing business this way. Then I asked Cheryl to read printouts of our correspondence. She picked out the verbal infelicities right away. "'Best four-wheel?'" she asked me with a crooked grin.

"You can't expect the sort of person who rides ATVs to write in proper English," I said in the seller's defense. This nice woman from Waterbury, recently divorced.

"Maybe not—but how do you know she's not in Poland? How do you even know you're dealing with a she?"

"Why would someone lie like that?"

"Honey, you're on Craigslist."

So that kind of put the kibosh on the Polaris 550 with the snowplow and the winch. A few days later, searching for ATVs again on Craigslist, I saw the exact same ad for the same machine. Same photos, too. But this time it was located in Bennington, not Waterbury. Two days later, it was being sold from Putney. Clearly some kind of scam was afoot here. Or maybe several of them. Possibly the ATV did not even exist, except as a gallery of purloined images; it was a con, like that proverbial letter from Nigeria. And I had actually considered sending $2,700 to a numbered account at eBay. Whoa. Buying a four-wheeler from a licensed dealer might be way expensive, but at least I'd wind up owning a machine. Buying one from Craigslist was a crapshoot, or a sucker bet. And I'd come *that close* to being taken in.

As winter turned to spring, then, I gradually despaired of acquiring an ATV to help me with my forest work. That was too bad, because once I started felling trees to open up the hickories there would be many cords of firewood to handle. And the trees I'd have to fell were nowhere near the forest's edge. But once I got that done the bats would have their new digs—their enhanced habitat—and I'd be released from working for the government. The pressing question, though, was whether there'd be any bats to find what I had done for them.

I tried to keep up to date on the bats' plight, but it was hard to get objective information. Articles in the environmental press tended to suggest that the sky was falling; articles by members of the caving community minimized the loss of bats and challenged the more sensational body counts. But scientists were hard at work on the problem, and there were reasons not to give in to despair. The specific fungus widely presumed to be causing white-nose syndrome had recently been identified; its genome had been sequenced, and it now had a Latin name. *Geomyces destructans* seemed apt and appropriate. It was now also known that although the fungus was most often seen on the noses of affected bats, more critical damage was being sustained by the thin, tough membrane of their wings—a membrane that accounts for four-fifths of a bat's overall body surface. And a membrane that, if compromised, would mean that a bat could hardly fly. On the bright side, it was turning out that not all species of bats were equally affected; there were hibernacula where one kind of bat had been virtually wiped out, while others were still able to emerge from their winter's rest in seemingly good health.

Then, in April of 2011, a fascinating paper on European bats was published in a scientific journal called *PLoS ONE*. Coauthored by twenty-seven researchers from all across the Continent, the article concluded that G. *destructans* was commonly found in European bat caves and on European bats—*without* causing any of the mass mortality that was unfolding here. The fungus was endemic on the walls of European bat caves; you could run a cotton swab across a cave wall and use it to get G. *destructans* growing in a Petri dish. Over the course of a winter's hibernation, the chances that a European bat would show signs of the fungus gradually increased. But the symptoms came and went. Hibernating bats were presumed to rouse themselves from slumber every couple weeks or so, and when they did they'd do some grooming to keep the fungus's growth in check. Then when they were ready to emerge in the spring, they'd do an especially thorough job of cleaning up.

One of the challenging suggestions of this research was that European bats might have long since acquired an immunity to white-nose syndrome—precisely what might be expected to happen here, given enough time for evolution to do its thing. And the European work gave rise to two scenarios: either G. *destructans* had only recently arrived in North America—and hence had blindsided our native Chiroptera—or else the fungus had already been here, but had lately mutated into something lethal. The former possibility seemed more likely, particularly since the first report of white-nose syndrome had come from Howe Caverns—a tourist attraction that is visited by thousands of people each year. Including Europeans, some of whom might have been spelunking back in France or Spain and brought the fungus on their shoes. That might have set this catastrophe in motion. Bats in a summer maternity colony come from many different caves, and get to share their cooties. Once G. *destructans* had arrived in Howe Caverns, the bats may well have spread it by themselves.

Seeing G. *destructans* as being of European origin came attached with policy implications. If the fungus could be deemed yet one more invasive species, federal money and scientific expertise could be marshaled to address it. Scores of different plant species were now held to be invasive; birds and mammals could receive that designation, too. Ditto for certain kinds of fish and snails and insects. Why not a fungus? Just as federal dollars could be spent on having guys like me poison buckthorn in the woods, federal dollars could be spent on battling G. *destructans*. So, in a way, that could be read as cause for optimism. On the other hand, by the coming of spring in 2011 researchers were in rough agreement that at least a million bats had already died. And apart from closing off the bat caves to human traffic—much to the chagrin of the spelunking community—no real strategy to fight the fungus had been found. Managing an outbreak of disease in a wild species is never easy; bats, for example, can travel far and wide and nobody can stop them. Anywhere they go, they're apt to spread what has infected them. And to eradicate a fungus that can grow in caves is next to impossible. Caves are dark and damp and cold; caves branch out in ways that make them hard to fully penetrate. If living on a cave wall is a fungus's niche, chances are it's going to beat even a well-funded effort to eradicate it.

The situation called to mind the "Spanish flu" epidemic of 1918–1919. My father had mentioned it in passing in his autobiographical document; on a particular day in 1919—when my dad was three years old—he and his one-year-old brother were ushered into their grandparents' bedroom to say goodbye. Not a wise idea, with benefit of hindsight, but grandma and grandpa both were ill with influenza. And they both succumbed to it before the day was over. Worldwide, fifty to one hundred million people died and five hundred million—or one-quarter of humanity—are reckoned to have been made sick by the virus. That's what we were up against, perhaps, with G. *destructans* and the future of the bats. But if six percent of human beings died of Spanish flu, ninety-four percent had not. Our species was still a going concern. And maybe bats would fare as well, once the dust had settled.

My father died of cancer, not some rogue and fatal flu. At the age of ninety-two—courtesy of Medicare—he'd undergone an operation to remove a kidney. He was of two minds about going through with it, but when the chips were down he chose to live a while longer. He was much diminished, though, after the operation. He fought on for one more year, but the cancer kept showing up in more internal organs. When it was concluded—by his doctors, at any rate—that his case was hopeless, my dad agreed to go onto hospice care. And when the end was near I went to spend a few days with him at the apartment where my parents lived outside Chicago. My father was angry; he could not believe that he would die. He planned to exercise and gradually regain his strength. Mind over matter, as usual. He could do it.

I didn't try to discourage this plan of action, but I couldn't help but see that Dad was taking constant risks. It wasn't ever safe to leave him unaccompanied. When no one was looking, he would reach to pull a heavy object down from a high shelf. Or disconnect a smoke alarm to change its battery. He had a walker, but he'd try to do without it and go teetering across the room. Once he would have fallen on his face, but wound up in my lap. Sometimes he would glare at me and then accuse me of one ridiculous charge or another—breaking a much-loved clock, or not speaking in a loud and clear voice, or settling for a job in life that he felt was far too easy and way overpaid. The last night I spent with him—the last night I would ever spend—he chose to fuss with files on his personal computer rather than engage in any human interaction. Next day, my mother staged a chance for us to say goodbye; my dad took an extra half hour in the bathroom, coming out only when I had to catch a train. It wasn't just about me, I realized. He didn't want to say goodbye because to do so meant admitting he was dying—or to admit that *I* suspected he was dying. So it was a botched parting, awkward and unfulfilled.

Within two weeks, his demise had become imminent. My younger sister, Mary, was on hand to help support my mother, and I called them one night after teaching an evening class. I spoke to my dad for awhile, but he couldn't hear me and he wasn't talking in a way that made sense. I said to my sister: "Look, nothing in Dad's nature has prepared him to see dying as anything but a personal failure. Someone has to tell him that it's alright to let go." After I got off the phone, Mary talked it over with my mother and that's what they did. They told him that he was allowed to die, if he wanted. They said it would be okay. And a couple hours later, my dad took his last breath. As if he'd been hanging on until he had permission. Next morning, in the barn, the year's first lambs were born. My father's niche was empty, and yet life was going on.

The death of a million bats had not been suspended pending anyone's permission; their grip on existence was not as tenacious as my father's, nor as self-willed. But for the bats, there was also a good chance that life would keep going on. It was a mistake, perhaps, to think of them as being discrete individuals—borrowing from the way we think about ourselves. Maybe the entire species was the operative unit. If so, then it was too soon to count them out. European bats had somehow learned to cope with G. *destructans*; bats in North America might learn to do so, too. But a sense of helplessness was hard to fend off, at times. Much had been learned about the causes of white-nose syndrome, but no course of action to combat it was in sight. What I was doing was as valuable as anything, working for essentially selfish reasons to improve bat habitat in the forests on our farm. *Warm season* habitat, which is when their pups were born—so it could be seen as crucial to the bats' recovery. But that was disconnected from the mass carnage going on in winter caves. Just as the crop of lambs we now had to deal with felt disconnected from the grieving process for my dad.

The grieving process went on for long after that year's lamb crop—and the next year's lamb crop, too. Not that I felt paralyzed or kept bursting into tears, but something about my father's death triggered feelings of personal impotence. That was why I'd built a lavish house addition, possibly. That was perhaps why I had tackled working in the woods and bought myself a chainsaw. He had made a point of absorbing pain without a whimper; stoicism on that order was beyond my ken. Every night that I came home complaining of aches and pains proved that I was not like him. My father had stood for a version of manhood that clearly exceeded mine, a toughness to which I felt unable to measure up. Without that impossible standard of achievement, maybe I was feeling lost.

Then, too, by the spring of 2011—two years after my father had died—there were other reasons for a sense of disempowerment. I had gone into the woods, but mainly as a low-paid weeder working for a government whose program I distrusted. I was still months away from getting WHIP's permission to start daylighting the roost trees so that bats could move into them. But if I did get the green light, that would be a monumental project for a man my age and with the tools I had to work with. That would be a ballbuster, even if lightning struck and suddenly I took possession of my father's fortitude. No, though. Most unlikely. What I wanted was a tool to make the job go easier. By the time that exercise in forestry was on my plate, I swore that I'd be working with an all-terrain vehicle. That, too, was arguably proof that I felt impotent. But in the greater scheme of things, it seemed essential.

By scaling back my original desiderata, I found I could get into a new four-wheeler without breaking the bank. Five hundred cc's was a good-sized machine for hauling out firewood, but a smaller engine only meant I couldn't pull as much. So I'd take smaller loads—six hundred pounds, say, rather than a thousand. I would have to make more trips, but at least I would be driving a machine. Kawasaki had a 300cc model called the Brute Force, selling for around $4,000. Yamaha was touting a Grizzly for the same money. I was getting close to choosing one of these machines when I made a final check on Craigslist, wary of getting scammed by con artists again. And there it was: an ad so terse and taciturn, it had to be for real. Along with a phone number I could call to actually talk with the seller. "2005 Polaris Sportsman 800cc like new $3,500 firm" read the copy. That was it. Eight hundred cc's—half the size of my canary-colored Porsche. I picked up the phone and chatted with a brusque-sounding man who said his name was Bruce, living near Rutland. He wouldn't give his last name on the telephone, nor would he disclose his address; he knew the ground rules for Craigslist transactions, and I was coming to appreciate them too.

"Look," I said, "I'm interested. But I want to see it first. Then, if everything's okay, I'll come back with the money."

"Fine," he told me. "Just so you know I'm only taking cash."

"*Cash*?" That was a deal-breaker. No way was I going to show up with that kind of money in cash.

"Or a bank cashier's check."

That sounded doable, although I realized that Bruce could grab the check and run. Lots of bad things could happen with a sale like this—sales between parties who don't know or trust each other, and who linked up via cyberspace. "Are you going to tell me where you live?" I asked politely.

"No. You pick a time and place to meet in Rutland, and I'll be there. Then you can follow me here."

"Fair enough. How about ten a.m. tomorrow, in the parking lot of Burger King?"

"Burger King. You got it."

"I'll be in a red Ford pickup," I said. "How will I know you?"

"Because I'll see you in your truck."

Rutland is about an hour's drive south of here, a straight shot down U.S. 7. Built on a vein of marble that has slowly been depleted, it's become a rough and financially distressed town. There are gangs in Rutland, and hard drugs like OxyContin on the streets. Burger King is right in the center of things, and its parking lot was all but empty when I pulled in a few minutes before ten. There was a man with a jumbo cup of coffee climbing into a Jeep; he looked up at me and we made eye contact. Did he also nod his head? Honestly, I thought so. But before I had a chance to ask if he was Bruce, he started up his engine and pulled out of the parking lot. *That's a little cold,* I thought. But I hurried up to follow, and he led me on a merry chase for several miles. Changing lanes willy-nilly, making hard lefts as a stoplight would be turning red. I stayed right behind him, showing I was man enough to drive with some authority. Macho. Quick-witted. Reckless. Showing him, I hoped, that I was worthy of his ATV.

Finally he turned into an ordinary driveway in a working-class neighborhood, hopped out and went into his house without a glance at me. I pulled up behind him, waving a printout of the ad I'd found on Craigslist. "Bruce?" I called to where I'd seen him just before the door slammed shut. Then I went to stand on his back porch, waiting.

After two minutes he came out, and he wasn't friendly. "What are you doing here?" he growled.

"Bruce?"

"There's no Bruce here."

"You're not selling a Polaris ATV?"

"The hell?"

"Burger King at ten. You said—I thought you said—" I pointed to where I had written "Bruce" and "Burger King" on the computer printout. "This isn't you?"

"Get out of here," he said. "Before I call the cops."

It was still like thirty seconds before ten o'clock. I backed out of there and raced back to Burger King, heart pounding as the adrenaline kicked in. I had just tailed a complete stranger to his house. In Rutland, where things get rough. After taking cover, he might have come to greet me with a hunting knife—or a Glock. Why should I have thought that a stranger getting in his car had to be this Bruce person? How could I have followed him without having exchanged a word? That was whack. That was batshit. True, buying an ATV had to involve a projection of manhood—and projecting manhood was much on my mind. But I had gone too far. All things considered, I was lucky to be alive.

Back at the Burger King, Bruce—the *real* one—was waiting impatiently in a mud-splashed Chevy. "Sorry to be late," I said. "I made a wrong turn. So—where are we going?"

"Halfway to Killington." Bruce didn't look at all like a skier type. He was a big man in his forties with a hunter's eyes—eyes that could draw a bead on something and pull the trigger. He had an air of dishevelment, too. "It's a ways outside of town," he said. "But follow me."

I followed him on a series of back roads that went straight uphill for close to ten miles; then we pulled into an isolated driveway with a tattered For Sale sign beside a rundown dwelling. I was feeling mighty glad I hadn't brought the money—either cash or bank check—since for him to take it off my hands would have been easy. Once again my heart was pounding. But we stepped out of our respective vehicles and I followed Bruce past a falling-down garage. And there it was, a black hulk parked beside a pile of snow. Beefy-looking. Rugged. Tough. I recognized it right away as my Batmobile. "Nice," I said. "As advertised."

"This was built in 2005," he told me. "The year Polaris went to electronic fuel injection. It's all paid for, and the paperwork's right here." He pulled a certificate of title from his pocket, showing that a bank lien had been recently discharged. "Not even four hundred miles on it, yet. Like new." Now he climbed onto the saddle, started up the engine and showed me the official odometer reading. Three hundred eighty-seven miles—that was nothing. And it had a winch, too.

He let me pull the dipstick, and the engine oil was clean. I gave each front wheel a shake; the bearings felt tight. The tires didn't show a lick of wear—but why should they, after so few miles? I figured there had to be something wrong with the machine, but it wasn't obvious. "Why are you selling this?" I asked him in a friendly way.

"Divorce. You been through one of those?"

"No," I said. "I haven't."

"Third time around, for me. Goddamn women, huh? Lawyers aren't cheap, either— that's for damn sure. Anyway, you want to take this out for a ride?"

I looked around. "On the road?" There was no place else that it could go.

"Sure. You're in the boonies, here. No one's going to bust you."

I shook my head. "I've never driven a four-wheeler."

He gave me an incredulous look. "Never?"

I explained that I had work to tackle in the woods, and that I'd heard a four-wheeler was the way to go. The tool of choice. What I needed.

"Oh, you'll do fine with this. It can pull a half ton, easy. Maybe more like fifteen hundred. I bought it for hunting bear—been to hell and back, I'll tell you. You'll do fine with it."

Now I noticed that the rear deck had what looked to be a gun rack. "Did you get yourself a bear?"

"I got more than one. But you can take the rack off, easy. Slap on a chainsaw box."

"Polaris makes those?"

"Polaris makes all kinds of ATV accessories."

At this point I let him show me how to put it into gear, and I drove for maybe twenty yards at five miles an hour. Then I shifted into reverse and backed up. That was it—my first ride. I turned the engine off. "Okay," I said. "Thirty-five hundred."

We agreed that I'd come back the next day with a bank check; then he had to give me his full name, since I'd need to have a cashier make it out to him. We took a tape measure and confirmed the ATV would fit into my truck bed. It would, but just barely—and only after taking off the chrome-plated ball hitch. Then he checked to make sure he had some lumber he could make a ramp with, when it came time to load the machine the following day. Everything was clicking. We shook hands and I drove off, feeling like a man. I seemed to sit a little taller behind the wheel. I threw my shoulders back, and noticed I could not stop smiling. I was going to be the proud owner of an ATV.

Next day, I swapped Bruce a bank cashier's check for the state's official certificate of title. Since I had no expertise in driving the machine yet, Bruce undertook to gun it up his homemade ramp into the bed of my truck. But his "ramp" turned out to be a short pair of two-by-sixes, laid in place up against the tailgate by eyeball. The angle of attack seemed impossibly steep, and the tires on the ATV were several inches wider than the planks they had to follow. There was a heart-stopping moment when the right rear wheel slipped off its rail, but somehow Bruce got the Batmobile on the truck. Four-wheel drive, I guess. And grim determination. I tied it into place with a length of heavy rope, and on the way home I stopped at a lumber yard and bought a pair of two-by-twelves. Ten feet long, too, so that the angle of attack on *my* ramp would not be suicidal.

Back at the farm, I parked the truck near an embankment of earth and managed to successfully unload the ATV. Then I found Cheryl in the kitchen, making lunch. "Hey, babe," I offered with a certain macho swagger. "Go for a ride?"

With Cheryl's arms around my waist, I nudged the shifter into gear and we took off for the woods. Not much faster than we probably could have walked there, but I was still getting used to the machine. It wasn't like our hair was flying in the breeze, but it was still an unforgettable ride. Up and down the various paths in the bat zones, where I had invested weeks and months of dogged effort. Then an electronic warning started flashing on the digital console: *HOT! HOT! HOT!* For a brief, happy moment I imagined that the message was referring to me. Riding on my Batmobile, rebel with a fine cause—and with a motorcycle mama on the seat behind me. Then I realized the words referred to the engine.

I put a stop to our ride and turned the engine off; yes, it *did* smell overheated. I found the radiator cap, but knew better than to mess with it until the machine had cooled down. There was a plastic expansion chamber, too—a see-through reservoir for topping off the coolant. And it was bone dry. So we walked back to the house, a bit deflated. But I came back with some water and a can of Stop Leak, on the hunch that Bruce might have dinged the radiator hunting for bear. And that fixed it. There has been no further trouble.

Soon the snow was melting as the long, hard winter turned to spring. My skill as a four-wheeler jockey grew by leaps and bounds. By May, I was zipping back and forth to the woods at thirty miles an hour—though I had to slow down once I got in among the trees. First I bought a chainsaw box to hold my logging tools, and then I found a sturdy cart matched to the Batmobile's size and strength. I could drive the full rig deep into the woods, then haul out firewood in satisfying loads. The cords of fuel were piling up: I had a woodpile that would likely last for three years, and the woods themselves were getting steadily cleaned up. New saw, newish ATV, free exercise. What was not to like?

Spring turned to summer, and we moved into the rhythms of gardening and haymaking and managing pastures. The Batmobile proved itself helpful with those tasks, too. I even built a twenty-foot bridge, allowing the machine to cross a gentle stream that drains the beaver swamp and get into the forests on the far side of the pond. There was another crop of weeds growing in the woods—garlic mustard, baby buckthorn—but I wasn't ready to attack that problem yet. There would be time, once the major farm work was completed. And the second year of this "invasives control" couldn't possibly be as hard as the first had been. So I let it ride all through June. And through July, too.

Then, the first week in August, I got a letter on USDA stationery with the subject heading "Annual Practice Reminder." The letter pointed out my legal obligation to get started on the second year of pest management in re: WHIP Contract #721544100Z1. Simply put, it told me I had better start pulling weeds again if I wanted to avoid a charge of noncompliance. The letter was more than a friendly heads-up; I took it as a wakeup call. And I was surprised to see the name of the person who had sent it to me: "Katherine Teale, Soil Conservationist." The return address was the same county office where, in the past, I had always dealt with George Tucker. I had never heard of Katherine Teale, but she seemed to think it was her job to check up on me. That was mildly annoying, though in truth I know that I needed a nudge. *Helping People Help the Land* read the printed slogan at the bottom of the stationery. Right, I thought. Sure. As if. I made a date to go to town and meet Ms. Teale.

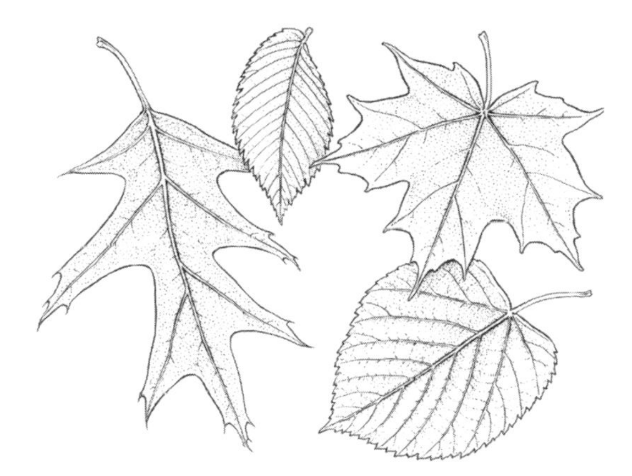

Echolocations

A couple days later, I stopped by the county agricultural office. Right away I ran into a harried-looking George Tucker, who told me that he'd recently received a promotion and thus was no longer responsible for me. Or for my contract with WHIP, at any rate. He walked down a corridor to summon his replacement, and a friendly young woman came to meet me at the counter. "Katherine Teale?" I asked, holding up my Practice Reminder letter. The one that bore her signature.

She smiled. "I go by Kate." She looked *awfully* young; she could have passed as one of my college students back when I'd been teaching environmental literature. At the very outside, she was in her midtwenties—which would make her several years younger than my daughter. And she was attractive, too: lithe, long-limbed, and impressively athletic. My, I thought. The face of stern authority had changed. This was Kate's first real job, in all likelihood, and I had a feeling that she wouldn't be comfortable telling a man in his sixties what to do. Not yet, at any rate. That might change after some time and experience, but for now I figured I would have an easy ride.

Kate dug my file out and looked over the documents. "So I see you're working on invasives," she said neutrally. It didn't sound like she was cheering me on.

"Only so you'll let me open up the forest later. The point is to daylight some shagbark hickories, so bats will want to roost in them. It's all there in the contract." I explained about the habitat we were developing, and how Toby Alexander said I'd have to fight invasives before I could be turned loose to cut down trees. "So," I said, "I've got a few more weeks of crawling through the woods, working for the government."

"Working for the taxpayer," Kate corrected me.

"Same thing, isn't it?"

"No. You're getting paid the taxpayers' money. It comes out of real peoples' pockets. Not some abstract entity."

Wow, I thought. I figured she'd been taught that line in WHIP school. But the bat dimension of our project seemed to interest her. "Thank Brendan Weiner," I said. "He's the guy who dreamed this up. Over at Vermont Family Forests—have you heard of them?"

"Sure. I've heard of Brendan, too—but I can't say we've met." Now Kate rummaged in my folder. "So did he draw up your Forest Management Plan?"

"Yes—and he was really thorough. Did a lot of research, too."

"I'm sure you're going to miss him."

"Miss him?" I shook my head, alarmed. Dumbfounded, too. Had there been an accident? Had a tree fallen on him?

"I was told that Brendan's left the state. That he's gone out west. Oregon, I think. Or maybe Idaho."

"When was this?"

"Just recently."

I must have smacked my forehead. "No one's said a word to me. You're sure about this?"

Kate look dismayed to be the bearer of bad tidings. "I gather it was sudden," she said. "He might still be around, but I doubt it."

Damn, I thought. If this was true, it could be a real setback. Brendan had been my agent in negotiations with the Landowner Incentive Program—which had paid for most of our Forest Plan—and with the people in charge of the state's Use Value Appraisal program. Not to mention dealings with the state's county forester—Chris Olson, with his office right down the hall—and the various honchos from the Natural Resources Conservation Service. No one else on the Vermont Family Forests team had the same level of investment in our Forest Plan; no one else knew the ins and outs of our situation. So while I'd been minding my own business on the farm, I had lost my forester—as well as the federal agent who had signed our WHIP contract and had been in charge of its oversight. How was I to do my part in saving the bats if these professionals kept playing musical chairs? "I can't believe nobody told me," I said, dismayed. "Brendan was a smart guy. He seemed to know his stuff."

"I'm sure they'll find someone good to replace him." Now Kate tapped the folder that lay open between us. "Anyway, you're saying this is still an active contract."

"Yes. As far as I'm concerned, everything's on schedule."

"Good. I ought to tell you, we're rethinking these 'invasives' contracts."

"What's to rethink?" I asked. Then I remembered some fine print from the year before about how the government could change a contract, if it wanted. It was not like I could, though. "I thought you guys believed invasives were important."

Certainly they are, she said. But in some instances the money being spent on them had been called into question. Sometimes landowners had been paid to control invasives, but the plants were *not* controlled—or not nearly well enough. A few years after a contract's completion, buckthorn or whatever would have moved back in. Thicker than before, even. In effect, some people had been wasting the government's money. Whoops—the taxpayers' money.

"But I'm not that kind of guy," I told her. "And besides, I'm only doing this because you told me to. Not you specifically, but WHIP."

"I understand that."

"And I'm really on the case. Ask George—he came out last fall. He said I was doing fine."

"All I'm saying is, don't count on getting more 'invasives' contracts after this one's finished. The program might have changed by then."

"Not a problem," I assured her. "I don't want another 'invasives' contract. Ever."

Kate laughed. "So—when do you want me to come certify?"

"Certify?"

"That means inspect it. Verify the work's been done."

"Oh." I scratched my head. "October?"

"That's a couple months from now."

I explained about the buckthorn and the glyphosate treatment—how it only worked in the fall of the year, when freshly cut stumps would suck the herbicide into the roots. I said I'd get started on the garlic mustard right away, but it made no sense to have her come out to "certify" before I'd killed the buckthorn.

"See you in October, then. Good luck with everything." We shook hands and I was on my way, relieved to know that I was still in an active contract. Even if the value of the war that I was waging was now under some kind of policy review.

Back home, I checked in with Vermont Family Forests and learned that Brendan Weiner had indeed departed—to Bozeman, Montana. His wife had gotten into a graduate program there, and they were unwilling to lead separate lives for the time it would to take her to complete a degree. They were, after all, raising two kids together. The parting from Vermont Family Forests had been amicable; there were hopes that Brendan might return, once this stint was over. But in the meantime he was out of the picture.

After meeting with Kate Teale, I began setting aside a couple hours a day to pull up garlic mustard—starting with the second-year plants with their *siliques* and their ominous rows of seeds. I had to get those into tightly sealed plastic bags, or else the party would go on a whole lot longer. In principle, there shouldn't have *been* any second-year plants if I'd done a thorough job the year before. But maybe I had overlooked some well-hidden whorls, or pulled them from the ground without removing the entire root. Anyway, the second-year plants were alive and well; there was no way to deny their existence. It was a race against time, I realized—but after winning it, a clean-up of the first-year plants looked to be vastly easier than before. There were still plenty of them, but they were more scattered around than bunched together in exasperating clumps. That, at least, was testament to my work on hands and knees.

Cheryl had been busy, in the preceding couple years, developing the concept of using the farm as a Retreat and Learning Center. Many years before, we had named the place after the surname that my father had relinquished as a boy. Treleven Farm, then. It seemed a way of honoring a heritage that otherwise would have been forgotten; after all, my own name would be Donald Treleven if my father's stepdad hadn't made him change his name. And so the nonprofit that we had incorporated was given a legal name that borrowed from the farm's name—and, in turn, from that of my father's lineage. Every couple months or so, we'd have a day-long seminar or conference here under the aegis of "Treleven, Inc." Often these would take place in the new Annex that I'd built alongside our house, but we always tried to incorporate an outdoor project. Something tied in with whatever work was currently taking place on the farm. So now Cheryl, sympathetic to my efforts to weed the woods, came up with a bright idea—and one consistent with her longstanding interest in contemplative practices. "We could have a Silent Fasting Retreat," she told me. "Spend a whole day pulling garlic mustard, while we meditate."

It sounded low-impact, that was for sure. We'd invite a couple dozen people to the farm, but we wouldn't feed them and they wouldn't be allowed to speak. All they'd be allowed to do was pull garlic mustard plants. Maybe we could even charge them something—ten dollars, say?—for the privilege of helping us weed the woods. If we pitched the project in the right sort of way, plenty of New Age types would likely sign up. *That* was Tom Sawyer and the white picket fence. We could root out all the garlic mustard in a single day, and everyone would go home happy. Grateful to have had a chance to do something to help the bats.

So we set a date—it must have been around the end of August—and began casting about for volunteers. There were Quaker circles to approach, since Cheryl is a member of the local Friends Meeting. There were circles of yoga practitioners and students of tai chi. There were child-welfare circles, school alumni circles, and our respective sets of personal friends. We even put a notice in the local newspaper, on their weekly Calendar. "Treleven, Inc." had a website that we felt was getting reasonable traffic; we were sure that we had done enough to get the word out. On the chosen day, though—a day that was fine and sunny, not too hot and not too cold—we waited for our driveway to start filling up with cars. Cheryl worked the telephone, accepting regrets from various people who had said they might be coming but whose plans had now changed. There were quite a lot of no-shows. Virtually everyone. By ten o'clock, it looked as if the Silent Fasting Retreat was going to be a total bust. But then one of Cheryl's colleagues from state government arrived, having driven over sixty miles to get here. That was it: a trio consisting of my wife, myself, and a cheerful woman from Montpelier named Brenda Bean. So we each grabbed plastic bags and headed to the woods.

Weeding is an exercise in pattern recognition, and it takes a while to teach someone the basics of the pattern. Especially if you're committed, for some zany reason, to not using words. So it was a game of pointing: leaves of this particular shape and color—these, not those. Flowers that should look like this. And don't forget the S-shaped root. Once a person starts to get the hang of a pattern, though, their eyes become effective filters. All the world divides into this one entity that you're trying hard to see, and everything else—all the things you're trying *not* to see. Cheryl and I used to do some mushrooming in these same woods, back in the day. We were young and not too cautious; we were finding out if we could live off the land. And it always struck me that at first you couldn't see a single mushroom, anywhere. Based on the season or the rainfall or whatever, we might be going after chanterelles, or shaggy manes, or—if we were really lucky—rumpled gray morels. Once your eyes had picked out one particular mushroom, though, suddenly you saw that they were scattered all around you. All across the forest floor. That's how it was with the garlic mustard, too, in that second year of weeding.

Teaching the pattern traits to Brenda—who was slow at first, but caught on quickly—got me to wondering why plants are so invested in their physical features. That was the focus of my meditative practice, whether it was appropriately spiritual or not. Having each specimen look just like all the others didn't seem that smart, from the garlic mustard's point of view. Why broadcast who and what you were with such fidelity? If they'd mix it up a bit, the plants that we were out to kill would have had a fighting chance. Some of them might have even slipped through our fingers. But they didn't do that, ever; it was as if they were all out advertising. "Here I am, a garlic mustard. Can you find me?" Yes, we could. Good thing for us—but did they *have* to always look the same? In the great scheme of things, what was the utility?

That got me to thinking about tree leaves, too. Nothing is more characteristic of a given tree than the shape its leaves assume; there's no mistaking an oak leaf for an elm's, say. Or a maple's, or a basswood's—and so it goes for each species in the forest. Why, though? What's the evolutionary payoff? Leaves have certain well-known functions, such as collecting sunlight and transpiring moisture. Leaves exchange certain gases with the surrounding air. Leaves are a locus for photosynthetic work. But none of these functions would seem to require that a tree's leaves all assume a characteristic shape, each and every one of them. It's a bit like hair coloration in humans: a prominent trait that seems to be without adaptive value. Black-haired people, brown-haired people, blondes, and redheads—from an evolutionary standpoint, what's the benefit? It was a distinction without a real difference. But how, I wondered, how can the trait that humans most employ to identify a given tree—to say that a tree belongs to one species, not another—how can such a prominent trait have no adaptive value? And if it really is without adaptive value, why are each tree's leaves so doggedly unvaried?

Well, that was food for thought. But then I started thinking about other human traits that seemed to be without adaptive value, and the distinctive sounds of voices came to mind. Maybe it was on account of how we couldn't talk, that day; we'd taken a vow of silence, and in consequence our personalities had been effaced. That was part of Cheryl's design for the retreat, though. *Having* a voice seemed a really useful adaptation; having a voice of a particular tone or timbre seemed to be beside the point. And yet human voices are enormously distinctive, almost like our fingerprints. (Fingerprints had no adaptive value either, one could argue, prior to the advent of detectives and police work.) Funny, though, how voices become complicated proxies for the person who lives behind them—even though none of us can actually hear the way our voices sound to other people. Hear the way we come across.

Then I started thinking, again, about my father's voice—firm and authoritative, laying down the law. That was a proxy for the person that he wished to be, or wanted to be seen as. But it wasn't really him. His voice was a device to mask the fact that he was insecure, and to guarantee that other people wouldn't notice. That's why he felt on solid ground by acting like the king. If he didn't talk to people in a certain way, he'd stutter. And once he started doing that, the jig was up. So his distinctive tone of voice *did* have adaptive value; it was a kind of armor, as well as a shrewd disguise. Maybe, I thought, every person's voice assumes those dual functions. Hence the sheer variety of human voiceprints—each of them tailor-made to fit a person's psyche.

I had been a stutterer, too, I now remembered. I couldn't pin down exactly when it started, but by first grade I was being taken out of class and sent into a small, dark room to get speech lessons. There had been a psychological evaluation, too, by a team of therapists at Northwestern University. That had been a memorable day—a trip to Evanston. Both my mom and dad were there; I had to wonder, now, whether my father had a chance to share *his* speech disorder with the child psychologists. Or if he did his best to keep it under wraps. As if he could do that—outwit the experts by talking at them in a firm, authoritative voice. Whether he did or didn't, he must have had a rougher time of it than I did. Anyway, they judged that I was not a deeply troubled child—not, at least, in psychological terms. I just had a hard time forming certain sounds and syllables.

Possibly my stutter came on because my father did it; that would have been a form of pattern recognition, too. Maybe it was even a hereditary trait. But by the time our family moved to New Jersey—when I was almost nine—the problem had been largely solved. *Not* by speaking in a firm, authoritative voice, though. For me, the answer was to learn a large vocabulary. That might seem bizarre, at first, but I'm pretty sure it's true. As their minds form sentences, stutterers can sense the word they know is going to trip them up. They have what amounts to radar for it; they can feel it coming from a clause or two away. Then you start panicking as the dreaded word approaches. When it's time to spit it out, sure enough—your voice screws up. But if you can plug in some alternative word—a synonym, or close enough—*that* word will come out fine. Not just single-word synonyms, either. I would learn to alter whole constructions on the fly. Actually, now I see that I was acquiring the skill set that a writer needs: phrasing and rephrasing thoughts until the words seem right. Doing this gave my voice an air of cogitation, which was better suited to my own personality than my father's urgent projection of authority. But we were engaged in the same project, he and I: trying to impose an artificial affect on our voices as a hedge against stuttering.

I still talk that way, I realized. I almost never stutter, but I live with a deep-seated fear that I am going to. If I let my guard down once. And I still edit my thoughts-in-progress constantly, sensing the word or words that might resist pronunciation and plugging in a substitution at the last second. Luckily, the English language has a huge stock of words; also, our syntactic and grammatical conventions are flexible enough that it is possible to say the same idea in many ways. A lot like bats using echolocation: setting off in one direction, then veering off course suddenly in ways that seem dangerous and hard to fathom—but without tumbling out of the sky. If you were inside the bat's head, it would all make sense. Dangers have to be avoided; opportunities emerge. Catching bugs at night without flying smack into things or getting nabbed by an owl must take quick thinking. So does giving hour-long lectures to college students without letting them once catch you tongue-tied. That was an exercise in echolocation, too.

At day's end, we broke our vow of silence and evaluated how things had gone. Nearly an acre of the woods had been rid of the offending garlic mustard, but we all agreed that the weeding had gone on too long. Maybe that's why so many volunteers had bailed on us. What we needed was more people, but for a shorter time. And the "Silent," "Fasting" dimensions of the day's design had not been really necessary; we could have meditated just as effectively and still be allowed to speak once in a while. And to eat some food, too—maybe even drink some wine. Maybe we were onto something, but the plan could use refining. Brenda got back into her car and set off for Montpelier; Cheryl sealed our plastic bags and put them where the next day's sun would bake their contents. "Treleven, Inc." had not exactly had a triumph, but we'd learned some things that ought to help us in the future.

Time kept marching on; Vermont summers go by quickly. Still, the work was going well. Once we got past Labor Day, I began spending most of each day in the woods. Half that time was spent mopping up the garlic mustard, and the other half was spent building trails for the Batmobile. As my network of paths kept expanding, I gained vehicular access to more and more deadfalls on the forest floor. Where these trees were not too far gone to serve as firewood, I would cut them up and haul them back to the house. As always, there was time to think—and my thoughts kept circling back to what I had been probing on the day of the retreat. My father's stutter, and my own, and how we each had engineered a personal solution. Then I recalled that my own problem had unfolded in the context of a famous book by Dr. Seuss: *Gerald McBoing-Boing*. I made a visit to the children's room of the local public library; the book was still in print after sixty-odd years. My parents had given me a copy circa 1950, when it first came out. It had been read out loud to me countless times, before I learned to read myself. The story is about a child born into a "normal" family who, as he develops, proves unable to form human words; every time he opens his mouth, sound effects emerge. But Gerald keeps on growing up, and soon he has to go to school where everyone makes fun of him. Then one day he makes a noise that frightens his own father, who turns on him angrily. On the verge of running away from home, Gerald is discovered by a radio talent scout and winds up with a job commanding money and respect; he becomes what today is called a Foley artist. A sound-effects man. This allows, eventually, a fond reunion with his parents, who are now proud of him. On the last page, Gerald's mom and dad go riding with him in a stretch limousine past admiring crowds of fans.

Well, now. It all came back. My parents must have felt that reading that book to me would help me feel better about speaking with a stutter. As if embracing it might vault me to some pinnacle of popular esteem. Or perhaps they felt that Gerald's plight was proof that speech disorders came much worse than mine; maybe they were hoping that Gerald's situation would put my own into perspective. Either way, it came across that Gerald was a character with whom I should identify. That we shared a common bond. Now, though—working in the woods, pulling garlic mustard—I couldn't help but wonder how I would have felt about my stutter if that book had not been current at the time. And it had been more than just an illustrated children's book; *Gerald McBoing-Boing* was produced as a cartoon that had won an Academy Award. Every child growing up around that time had either read the book or seen the film. Or, most likely, both. But for children like myself who really had a speech disorder, coming to grips with it was not made any easier by living in the shadow of this well-known caricature.

One night in mid-September, I did some research into the presumed psychology of childhood speech disorders—that's how deeply it was weighing on my mind. And for children whose stammering or stuttering had a sudden onset, one line of speculative thought came through loud and clear: sexual abuse, and particularly by an older relative whose standing in the family was above reproach. That honorific status made the sexual behavior difficult for the child's mind to process. Unable to parse where a line had been crossed between ordinary gestures of affection and something else—something more intimate—the child feels confused and frightened. And perhaps he or she even feels responsible for whatever has taken place. So the child enters a conspiracy of silence with the sexual aggressor; consequently, these situations can go on for years.

I filed that away, and took it to the woods with me. I did not regard myself as someone who'd been sexually abused as a child, but I had been challenged now to rethink my experiences. And I had the time to do it: I was pulling garlic mustard, piling up dead limbs and stacking runs of firewood. Work that let me choose my food for thought and chew it thoroughly. What might have happened to me? When had I been traumatized? Gradually it came to mind—and not in any blinding revelation, but rather in a slow and hard-won accretion of details. Memory by memory. *Yes, that really happened to me. Yes, that's what I remember. Yes, I'm sure that that took place.* And the memories all involved my father's stepdad. Gramps.

One random moment in this process of discovery came by sheerest chance, when I overheard somebody use a long-forgotten phrase. It was an electrician who I'd called in to solve an off-and-on wiring problem; we had a circuit breaker that would trip for no good reason, and I couldn't find out why. The man touched something in the load center that threw a spark, and as he jumped backwards he exclaimed "What the dickens?" Then he found a wire nut that had started going bad and swapped it for a new one. Everything was fine again. But the phrase he'd used stayed with me, teasing at my vault of memories. After a while, I was able to remember where I'd heard that phrase before. "The dickens" had been one of my Gramps's favorite expressions; he would use it in a wide variety of contexts. One of them, though, was this: "You've got the dickens in you." He used to tell me that with a merry laugh, squeezing my shoulder or tousling my hair. Had he thought his stepson's boy, at the age of four or five, would understand a reference to the writer Charles Dickens? Actually, I might have; my father liked to read *A Christmas Carol* to his kids out loud, and he was obsessive about crediting authorship. But one day I left my work unfinished in the woods and came home to look up what "the dickens" could have meant. "The dickens"—small *d*—is an Elizabethan term referring to the devil. That made perfect sense, because I now remembered Gramps saying that I had the devil in me, too.

My father's mother died in 1945, after yet another bout of major depression that had placed her—once again—in her mother's home and under Della's not-so-tender care. After that, my father's stepdad—whom we always called our Gramps—was freed from putting up with his wife's mental illness but became a sad and lonely man. By now he was on the tenured faculty at Madison, teaching agricultural economics; we'd drive up to see him as a family several times a year and spend a weekend filling up his dreary, cramped apartment. Early in the morning on the days when we were visiting, I was encouraged to enter Gramps's bedroom and arouse the man by pulling back the sheet and tickling his feet. That was always good fun. Then I'd climb in bed with him. Was my older sister with me? Yes, I think she was—sometimes. Anyway, Gramps's blue-gray eyes would open and he'd sit up to start playing a favorite game with me. The game was called "This Little Inch"—and sixty years later, pulling garlic mustard from my woodlot in Vermont, I found that I could summon forth each and every word of it. "Oh, you've got the dickens in you—yes, you do!" Gramps would begin in a jovial tone of voice. "I just love you to death, do you know that? I love every inch of you! I love *this* little inch"—and he'd grab, let's say, my ankle and give it a friendly squeeze. "And I love *this* little inch"—reaching to play with my kneecap, or my lower thigh. "And I even love *this* little inch!" At which point he would reach higher up and stroke what he found there. Loving every little inch. Am I here accusing Gramps of fondling my genitals? Honestly, I don't know if he did or if he didn't. But his way of touching me was highly inappropriate; surely today it would be legally actionable. I was not inclined to be litigious, though. I liked the game. He had introduced me to a secret, naughty pleasure.

Gramps had a television even before we did, although in the early fifties they were still a novelty. There couldn't have been more than two commercial stations in Madison, Wisconsin, but one of them would often broadcast wrestling on the weekends. Saturday afternoons, and even Sunday mornings. When I think back, wrestling seemed to be on constantly. In Madison, that is. To a small and unathletic boy, it was astonishing. Two well-built, half-naked men would circle round a ring, each plotting ways to take the other person down. Who would have thought this could be popular entertainment? But it was, in that place and at that time—and my Gramps seemed to be a wrestling addict. He took time to demonstrate the various holds to me—headlock, hammerlock, half nelson, scissors grip. After the contestants had grappled on a mat for a sufficient length of time, one of them would be in the power of the other and unable to move; that was called *submission*. I'm sure I was fascinated, but also surprised that my Gramps would be enthralled by such a violent sport. And such a lurid one. I'd been told many times that Gramps was a professor—whatever that meant to a child of four or five. Well, it meant a person who was really, really smart. And whose intelligence had been rewarded with some letters he could put behind his name. He had something that was called, in hushed tones, a PhD.

Then the day came when being left alone with my Gramps caused me to throw a fit. It must have been in autumn, because my mom and dad were going to take my older sister—Susan—to a UW football game. The Badgers. At their stadium, Camp Randall. I was crazy jealous, but for some reason it had been decided that I couldn't go. Maybe I was thought too young to understand the game, or maybe I was thought to be catching a cold. Maybe they had only been able to get three tickets. Anyway, I couldn't have been more than four years old— because if I'd been five, my younger sister Mary would have been there, too. And she wasn't on the scene. So—1951, then. I started freaking out even before my family ducked out of Gramps's apartment, but it wasn't just about being left behind. It was about being left behind with him. What were we going to do for a whole afternoon—watch more wrestling on his black-and-white TV? Play a three-hour game of "This Little Inch"? I remember going fetal on a sofa covered with a stiff, gray fabric. I remember crying so convulsively, I couldn't breathe. Blue in the face, no doubt. Gramps could not console me, and eventually he stopped trying. When my body couldn't shake or bawl any longer, I fell into a deep sleep and didn't wake until I heard my parents coming back from the game. Then I cried a whole lot more.

So something had been going on there, and it wasn't something I had learned about in Sunday School. *Oh, be careful, little hands, what you do*. I had all the time in the world to explore that history, picking garlic mustard plants and stuffing them in plastic bags. There it was: my Gramps had had a homoerotic bent, and he was a pedophile. He was a sexual predator who'd preyed on me—or at least had tried to. And how far did he get with that? Memory is layered, and there were certain layers beyond which I could not go. Why should I remember, though, his wiry tuft of pubic hair? Why should I remember the sharp stubble of his beard? Maybe, as he often said, because I had the devil in me. Or was it the dickens? Maybe I kept quiet about what was going on with him—whatever had been going on—because I felt responsible.

But in case I did, I had the shoe on the wrong foot. It's not right for grown-ups to make children into sexual playthings. No matter what.

Anaïs came home for a few days in late September, back from yet another busy round of touring. Her next CD was pretty much in the can, and the title song—like the album itself—was called *Young Man in America*. She had a wild, "rad" idea for the cover art: she wanted to use a photograph of me that was taken when I'd been a young—well, youngish—and defiant-looking man. Also somewhat stoned-looking, if you looked at one side of my face and not the other. The photograph had hung in a gallery on our kitchen wall for the last twenty years, framed and safely under glass; now my daughter asked to take it down and send it to the graphic designer who was packaging her album. "Fine with me," I told her. "If that's what you want to do." A lot of the songs, I knew, were about youthful rebellion and intergenerational conflict. And there I was, a rebellious young man. And some of the songs were an oblique examination of what she had seen in my dealings with my father—in our strained relationship. I knew these things, and I admired her courage in tackling such material. What's more, the photograph of me was taken when I had just turned thirty, and Anaïs was now thirty years old herself. So there were artistic connections on several levels. And it was a moving tribute, everything said and done. What singer-songwriter puts her father's portrait on the cover of her new CD?

There's nothing that I shy away from sharing with my kids, because as a child too much had gone unshared with me. So I told Anaïs about what I'd been discovering—or unearthing, really—from my past while working in the woods. She wasn't shocked like I'd been; she didn't even blink. Actually, she went a step beyond what I'd deduced. "Your father must have known," she told me. "I mean, think about it. How could he have not known that?"

"Known what, exactly?"

"That his stepdad liked to play with boys. Wouldn't your Gramps have tried to play with him, too?"

"Whoa," I said. "I just can't go there. I can't imagine it."

"Do you think Wayne's mother knew?"

"I never met my father's mother. She died before I was born."

"I don't know how something like that could be a secret."

But *I* knew. When families decide to avoid a subject—even if it's only a tacit agreement—that subject has a way of staying under wraps. Take sex, for example—in my family no one ever, ever talked about sex. Whatever I eventually managed to learn about it, I had to figure out on my own. Maybe there was some connection between that reticence and the things that people should have known were going on. Or maybe they didn't know, or even have suspicions. But the code of silence had a clear beneficiary, and it was my father's stepdad. It was my Gramps.

With the garlic mustard infestation now well in hand, I was able to go back to fighting glossy buckthorn. The younger, smaller plants that I'd uprooted the year before and burned to a crisp didn't magically come back to life; that was one survival trick they hadn't got the hang of, yet. But in the plots where I had yanked them from the ground, dozens of young, eager shoots were now emerging. Sallying forth to replace their fallen ancestors. Why should this be? The most likely cause was soil disturbance. When you pull a woody plant out by its roots, the litter of decaying leaves and forest trash is pushed aside and fresh soil gets exposed. Soil that tiny seeds have been lying there in hopes of finding. If you imagine that the forest floor is covered with zillions of seeds just waiting to germinate—which, in fact, it is—then exposing fresh soil gives them a shot at that. In a place where buckthorn berries had ripened on the branch and then fallen to the ground, giving them some soil to connect with must have made their day. Since the new crop of buckthorns weren't even one year old, they weren't hard at all to pull out. But there were a lot of them. This time, though, I made a point of tamping the loose soil back in place and covering it with leaves and litter. That might keep the next round of seeds from germinating.

It was a different story with the larger buckthorns that I'd cut with the chainsaw and then poisoned their stumps. They were all goners, and the soil hadn't been disturbed—so there was no rising generation coming up around them. Where I'd missed a stump or two with the glyphosate, though, a wig of green shoots would have sprouted to make fun of me. That was why The Nature Conservancy had recommended leaving an eight-inch stump poking from the ground. More than enough to trip on—and trip on them I did—but with that much stub to play with, I could cut it off again and hit it with the KleenUp. This time, for sure.

It took a week or so to finish off the buckthorn, but it was a matter of time and distance. Methodical exploration of the now-familiar bat zones for the umpteenth time. I kept thinking back to what Anaïs had suggested—that my father would have known it if his stepdad played with boys. That my dad himself might have been subjected to such play, and touched inappropriately. It made me feel creepy—and it also made me feel empathetic toward his mom. If my dad had known about it, wouldn't she have known it, too? But if she did know, then her life would have been hell. I looked up some vital statistics in the family tree; Lillian had died in July of 1945, at the age of fifty-one. Then I noticed that her death came six months after her second son, Earl—by her first husband, Guy Treleven—had been killed in World War II. The Battle of the Bulge. That must have unhinged the woman, given her already fragile state of mind. That was why she'd been sent to her mother's again. To Della. That's where she had been living when the end came for her. But if Lillian suspected what Anaïs did about her second husband, Donald Mitchell—my own namesake—then the end could not have come quickly enough.

On one of the last days of poisoning the buckthorn, a conversation with my father from thirty years before floated into consciousness and made me put the KleenUp down. It must have been in an unguarded moment, because he almost never talked about his family. But I remember we were driving in his car together, and somehow the topic turned to Lillian and her depression—what my father called "the mental merry-go-round." At some point, I asked him when and how his mother died. With a wry laugh, he told me no one really knew. But he suspected that *her* mother—my dad's grandmother, Della—might have done her in.

Good thing I wasn't driving, because the car would have probably gone off the road. No, I told him, that couldn't possibly be true. First of all, it would have been a case of murder. Charges would have had to be brought; there would have been a trial. My father was unfazed. He told me that the coroner's report gave the cause of death as being heart failure, but as far as he knew that was a medical *condition*—and not one that his mother had. Heart failure was not a plausible cause of death. People died of heart *attacks*, not heart failure. Anyway, Della had a mean streak and she could have done it. Maybe poison was involved. Possibly, he admitted, it had been a suicide. But he didn't put it past his Granny to have killed her daughter. Lillian's life had not turned out, and things weren't getting better. Maybe Granny felt she had the right to pull the plug on her. His tone of voice was wounded, but I wouldn't call it outraged. It was not condemnatory. It was, I now realized with a shiver, as if Lillian had been sent *to the Foot* in a game of "Who, Sir? Me, Sir?" A *Foot* you couldn't claw your way out of, after you'd been sent. A *Foot* that meant Game Over.

Wow, I now realized. The things that come at you when you open Pandora's box. In a way, whatever "Granny" had or had not done to her daughter—to Lillian Hollander, my father's mom—was totally beside the point. The point was that my father entertained the possibility. *Filicide*—that was the word for it in Latin. He believed that filicide had happened in his family. He never raised his suspicions with me again, though he must have lived with them each day of his life. Put it all together, and my father had grown up in a dysfunctional family. Colossally dysfunctional—and a house of lies. Teenage girl gets pregnant while she's working as a housemaid, so she marries her employer's son who's old enough to be her father. She loses that baby, but bears him two more sons amid recurring bouts of clinical depression. Then her husband dies and she remarries to a man who is perhaps more attracted to her boys than to herself. She has a third son by him, but now can hardly function. My father watches all this, and somehow lives through it. Coping with the mother who is mentally ill, and with the "clean" but twisted stepdad. Then his younger brother dies in battle, and his mother goes off the deep end. Then *her* mother takes it on herself to kill her daughter, and she somehow gets away with it. Bottom line: no matter what he thought about his upbringing, my father had not been adequately parented. In some ways, he had not been parented at all. What could he have known, then, about trying to raise *his* children? What could he have known about parenting me?

Everyone who sets out to raise kids finds they have to make decisions on the fly, without advance warning. When to be encouraging, when to be sympathetic, when to assume the role of disciplinarian. When to be generous with affection, when to hold it back. You make up what you're doing as you go along, because you *have* to; parenting is like bats prowling darkened skies, making their way by means of echolocation. To some extent, moms and dads are often flying blind. But I think my father had to do it more than most parents—and he had to do it with an air of authority, since that was the mask that he had taught himself to wear. And he wore it not because he *was* an authority, but because *acting* authoritative worked for him. It was essential to the niche that he had carved himself. Now, for the first time in a long time, I felt sorry for him. When it came to being a father and to raising kids, my dad was in waters way over his head. If he'd ever talked about his past, I might have understood that. But he had determined it would stay a closed book.

I finished up Year Two of controlling exotic invasives by the first week in October; this time around, it had taken only half as many hours as the year before. Good thing, too, since the contract's scheduled payment was only for half as much. One fine autumn day, I phoned Kate Teale at the Natural Resources Conservation Service office and asked her to come out and certify my work. We arranged a time to meet, and I primped for it like a young man going on a date. I had Cheryl give me a haircut; I put on clean clothes. I even hosed down the golf cart and washed the seat, since that would be the vehicle I'd use to chauffeur Kate to the woods. No riding double on the saddle of the ATV. And I made a last-minute tour of the bat zones. No fake problems for the government to find, this year. I wanted to impress Kate with a site where invasives had been perfectly controlled, giving the taxpayers value for their money.

Kate showed up right on time, and couldn't help admiring the contours of our private valley. The lay of our land. The variety of interesting natural communities—forests, meadows, pond, cliffs—as well as the beauty of the way the pieces fit together. "This is really something," she said. "How long have you lived here?"

I said that we'd bought the farm in 1972, and moved here two years later. "We were just kids," I told her. "We were twenty-six years old."

"*I'm* twenty-six," she told me. "And I've got a ways to go before buying a farm."

I explained—briefly—about my youthful hitchhiking novel and my stint as a screenwriter in Hollywood. Back in 1970. "Crazy, huh?" I asked.

"I guess."

"But life is a journey. Now I pick weeds for the gov—oops. The taxpayers."

We climbed on the golf cart and proceeded to the woods. Kate had another brand-new GPS device, and she wanted to walk the bat zones' lines and double-check the acreage as determined by George Tucker the year before. After ten minutes, though, she changed her mind on doing that. The long runs of baling twine clearly impressed her, not to mention the flagging tape wrapped around each shagbark hickory. From time to time we'd walk by stumps of poisoned glossy buckthorn, and she'd stop to photograph them with a digicam. Especially if there was a bunch of stumps together, it was good to take a picture. An "after" shot. Visual evidence of what had been accomplished. Other landowners in the district, she said, were working on invasives but without a higher purpose such as trying to help bats. Fighting the buckthorn with a less-specific goal in mind. Biodiversity, say. Or the overall quality of timber. At least it was fortunate that government programs were paying landowners to undertake such work.

"That's a hard way to make money," I objected. It was the hardest path to riches I had ever tried. Then a scary thing occurred. Walking down a path in the woods that had become familiar, there before us stood an unpicked stalk of garlic mustard. The second-year variety, with the tan *siliques*—although the seeds had now been shed. It was a desiccated ghost of its former self; it had lost its heart-shaped leaves, and its stem was brown and withered. Still, there it was in the middle of a bat zone, raining on my parade. How could I have overlooked it? I felt my pulse rising, and my face began to burn. "Oh, no," I said. "Shit."

"Something wrong?" asked Kate.

I stooped to dig the plant out of the ground. "I'm really sorry about this."

"What is it?"

"Garlic mustard. Not the first-year type—the second."

She looked at the plant with interest. "Are you sure?"

"I guess. I mean, I've picked quite a few of these. First-year garlic mustard grows low to the ground, in whorls. This is what they look like in the second year—their seeding stage. After dying back."

"Well," she said, "I only moved here just a short while ago. I know garlic mustard in its flowering stage, but I don't believe I've ever seen it like this. I'm still learning about some of your invasives."

"So you don't think this is one?"

"Not that I'm familiar with."

Well, now. Could it be that she was right and I was wrong? Maybe. I mean, it was certainly possible. But I was pretty sure that I was right and she was wrong. But we couldn't both be right about this, and we couldn't both be wrong. Then I thought about how Brendan Weiner had told me I had Norway maples, and George Tucker said I didn't. Both of them forestry professionals, too. Then I remembered what you had to do to honeysuckle bushes to see if they were native or not: break the central stem. Then I thought about the endless quibbling, a year ago, about which particular trees near a shagbark hickory ought to be removed to get more sunlight on the roost tree's trunk—but without severing its tendrils of connection to adjacent forest canopy. What we were involved in here was not an exact science—not by a long shot. Trained experts disagreed. And I was an amateur. Could it be that I had picked thousands of plants that were not, in fact, second-year specimens of garlic mustard? The thought made my head spin.

I scrunched the plant into a wad and stuffed it in my jeans. Then I raised the question as to who should come out and mark specific trees for me to fell when the time came to finally open up the woods. As I understood it, I could start this work November first—when it could be safely assumed that bats had decamped to find their winter hibernacula. "Would you know which trees to mark for cutting?" I asked Kate.

"Not yet—I'm new at this. You should get a wildlife biologist to do it."

"Brendan seemed to think he knew," I told her. "But he's gone, of course. What about David Brynn?"

"Why him?"

"He was Brendan's boss."

"That might be okay," she said. But I should consider someone working for the government. Toby Alexander, for example. Or George Tucker. Then she gave me someone else's name, and I wrote it down.

"How about Jane Lazorchak?" I asked her on a sly hunch. The hunch was that I might score points by plugging for a woman and thus showing I was gender neutral. "She's in charge of something called the Landowner Incentive Program—LIP—at Vermont Fish and Wildlife. And she's been out here before. She knows something about bats."

"Jane could maybe do it," said Kate. "It wouldn't hurt to ask."

"I still have her e-mail address, I think. Somewhere. But is this my call to make? I get to choose somebody?"

"Keep me in the loop," Kate told me. "I'll ask around, myself.

That's the way we left things. Once again, I'd passed inspection with my cleanup of the woods; Kate did find a couple buckthorns that I'd overlooked, and marked them with flagging tape so I could get them later. But it was no big deal. Overall, my work had been impressively thorough. There was no reason to deny my scheduled payment. Five hundred eighty dollars this time around—just a bit less than I had spent to buy a chainsaw. We got back in the golf cart and I drove her to the house.

"I'm sure we'll find somebody who's qualified to mark your trees," Kate assured me as she climbed into her truck. "And when we do, I'd like to be here."

"Why?"

"To see how it's done. We're all on a learning curve, with this sort of project. There haven't been too many bat contracts before."

So that was it. Two years of waging war against invasives, and the government believed I had them on the run. My work had been certified. In the days and weeks ahead I tracked Jane Lazorchak down and asked if she'd mark trees to cut, assuming she was qualified; meantime, I put in the same request to David Brynn. He said he *was* qualified, but also really busy. Then, too, if he did it he would have to charge me for his time. He suggested I find someone working for the government—that way, the bill would be picked up by the taxpayers. He said he thought that Jane Lazorchak would be fine.

But Jane wrote back a few days later, saying she had reservations; she had never marked trees for cutting before, and she wanted to check in with the state's Wildlife Management program director—Scott Darling—to see if he believed that she was suited to the job. A few days later, on my sixty-fourth birthday, Scott Darling e-mailed me. He did remember his night here several years before, and how he'd successfully trapped two Indiana bats. He'd heard about our contract with WHIP, and he was enthusiastic. Though he had a busy schedule, he wanted to come back and personally mark the trees he felt I should remove. Could we settle on a date?

"Wow!" I said to Cheryl. "We are going to get the bat man."

"The same guy who was here before?"

"The state's leading chiropterist." I was jubilant. First of all, Scott knew his stuff; he was now routinely being quoted in the press with news about the bat crisis. Second, his expertise was recognized by everybody in the several government agencies I had to please. Third, when it came to bats I figured I could learn more from Scott than from anyone. Chatting him up as we walked around the woods together would answer lots of questions. I could pick his brain and probe his special expertise. Then perhaps I'd finally understand what I was doing.

I checked in with Kate Teale as Scott and I wrangled to choose a date; October 27 seemed to work for all three of us. Then Kate let Toby Alexander know what was afoot. He wanted to drive down from Colchester and be on hand to see which trees Scott Darling would mark, too. So—another party at the taxpayers' expense. Fine with me, I said. Why not? The more the merrier. October 27 worked for Toby, so we made it official and I circled the date on our engagement calendar. Then I turned the page and put a circle on November 1. That was the date when I could finally go into the woods and daylight shagbark hickories. That was the day when I would finally go to bat.

Authority

The morning of October 27 broke cold and raw, with a stiff wind and a good likelihood of snow. That would not be problematic unless we got several inches—possible, although not probable so early in the season. But there was also the prospect of freezing rain, and that would prevent Darling's marking paint from sticking to the bark of trees. I kept waiting for the phone to ring and hear him cancel, but it didn't happen. At nine o'clock a state truck pulled into the driveway, and I went to meet the driver. "Right on time," I said.

"How you doing?" Scott nodded as if he recognized me, though in fact he couldn't have. "It's been a while," he said.

"I don't think we've ever met in person."

"No?"

"Just some phone calls."

"Then maybe I've forgotten. Place looks familiar, though." He stepped out of the truck and we shook hands; he was wearing an official green vest with the state Fish and Wildlife logo. No hat, no gloves. In his midfifties, roughly. Darling's eyes were at once observant and bemused, set behind spectacles and underneath a sloping brow. "When was I here before?"

"Five years ago, I think. Or maybe it was six. Before anyone had heard of white-nose syndrome."

"That sounds right." He opened the passenger door of the truck and rummaged on the floor to find a shiny metal can. It had an internal pumping mechanism, but when he squeezed the trigger nothing came out. Something was gummed up in there. "That was a good night's work," he told me. "Kept us busy." Then he used a wire brush to ream out the aperture; a bluish stream of paint emerged and landed on the ground. "Looks like we're in business."

"How are the bats doing?" I asked, trying not to show too keen an interest. I knew that reporters put this question to him all the time, and I was expecting a Cassandra-like response. That's the way I'd heard him come across on television, putting a pessimistic edge on his assessments. Here, though, standing in my driveway with no cameras rolling, he didn't seem to think the sky was falling. Little brown bats were in terrible shape, but he'd found some colonies on the far side of the state that seemed alive and well despite the general die-off. Understanding how they did that was the pressing question. Did they have some built-in resistance to the fungus—and if so, what could it be? Had they somehow dodged exposure—and if so, by what means? Or did they behave in ways that, despite exposure, somehow kept the white-nose syndrome from taking them down? Those were the three most likely hypotheses; now the trick was buying time enough to test them out.

Kate Teale pulled in and parked her truck alongside Scott's. Since she was new to the state, the two had never met; they had certain rituals of salutation to perform. Welcome to the team, and all. Welcome to the neighborhood. Kate said Toby might be coming down from Colchester, but she was no longer sure. His wife was nine months pregnant, and he might have decided to stay close to home. After waiting for a while, I began shivering; I ducked into the house to grab a parka and my winter gloves. No point in freezing, or in trying to suggest I was immune to bad weather. Then the three of us set off to start our tour of the woods, with Kate glancing back from time to time to look for Toby.

The first shagbark hickory tree was right along the forest's edge, and Scott quickly sized it up. It was a big, mature, canopy-layer tree—an obvious dominant—but its location made it almost too exposed to be attractive as a bat roost. I could sense the expert's disappointment, having driven up from Rutland to be shown this tree. I was disappointed, too, since it was the only shagbark hickory in this bat zone. I had rooted out half an acre of invasives to "protect" this one tree from buckthorn and garlic mustard, but now it turned out to have limited roost potential. Still, we were here to make it work as well as possible. Scott walked us through the first few stages of his calculus. First, you had to think about the sun's angle in midsummer. Then you had to see if nearby trees were going to block that light. Then you had to factor in desirable canopy in the tree's vicinity, allowing bats to fly in and out with a degree of safety. All these calculations, though, were made with an eye to how the woods would look in twenty years: which pole-size trees would have grown up to be dominants, which current dominants might have died and turned to snags. How the target tree itself—the shagbark hickory—might have grown in twenty years. It was like being asked to peer into the future.

There was something else, though, that he couldn't really share with us because it was too complicated. It was like an intuition that his work had cultivated. Scott could look up into the branches of the canopy—stripped of all their summer leaves—and he could imagine how a bat would move around up there. Where it would be likely to fly higher up, or swoop down. Where it would bank left, or right. Bats were little *acro*bats; they could move around in the medium of air the way a fish can move in water. He tried to explain how he decided where a bat would go, but we couldn't really follow. We hadn't spent our lives studying the ways of bats. Finally he took the paint gun and marked a couple trees—first at breast height and then right at ground level, so there would be evidence on every cut stump that the tree in question had been chosen for removal. Even in this relatively simple situation, Scott's choices were not entirely obvious. But they were authoritative once that paint was on the trees. When it came to planning better habitat for bats, he was the man. No one in Vermont, at least, could match his expertise.

We started making our way deeper into the woods, but then Kate heard Toby Alexander's truck arriving and she doubled back to meet him. Soon Toby had joined us, too. At first I was apprehensive; Toby was the main reason why I'd had to spend several months of my life crawling on my hands and knees. What if now he saw I'd missed a garlic mustard plant or two? But he greeted me with a friendly, boyish grin. As if I'd survived an ordeal of initiation, and could now be welcomed as a member of the tribe. Then I reminded myself that my WHIP checks were already deposited; he would have a hard time getting back the money. So I made nice with him, realizing that he'd come here not to check up on me but to learn from Darling. We were all Scott's students for the next several hours.

The roost trees seemed to get more and more interesting as we pushed into the forest. Scott showed us how there was considerable variation in the way the plates of bark hung off different hickories. Some of them would only be of passing interest to a bat; others were so richly exfoliated that the bats were likely to take notice—and, hopefully, move in. Sometimes Scott would walk up to a shagbark hickory, sight up its craggy stem and sigh in appreciation. "Oh, Mama!" he exclaimed on one or two occasions. A really good roost tree could house a maternity colony of several hundred bats at once, each of them nursing a pup beneath a plate of bark—and you might never even know that they were there. But even at the biggest and best-shingled shagbark hickories, Scott rarely marked more than two or three trees in the vicinity to be removed. Two or three, not five or ten. The forest's general character would not be greatly altered after I had cut them down. It was a much more conservative approach than had been talked about the day when Toby and George and Jane and Brendan had tramped these woods. In fact, after all of their concern about the downside risks of opening the bat zones, it looked like they weren't going to be opened much at all.

Talking shop with Darling, Toby asked if he was still moving ahead with something called "the bunker plan." This turned out to be a scheme to round up healthy bats and move them into unused military bunkers to spend the winter. The bunkers were in Maine or New Hampshire, or maybe both, relics from the Cold War or even World War II. They could be made to serve as surrogate caves or mines—the bats' traditional hibernacula—except that they came equipped with climate controls; by keeping down the level of humidity, growth of G. *destructans* could perhaps be inhibited. Yes, said Darling. That plan was still being pursued. Not for the coming winter, but perhaps the next. Data loggers had been installed in the bunkers to see how effectively the various controls would work. It was a dice roll, but bold measures were required. When it came to decimated species like *Myotis lucifugus*—the little brown bat—unless some cohort could be kept alive by one creative means or another, there was little reason to keep searching for a cure.

Amazingly, after several years of investigation the specific mechanism by which G. *destructans* killed the bats was still not fully understood. Maybe, though, it had to do with something as simple as disturbing their sleep. Like when your skin gets itchy, or you have eczema—who hasn't spent a sleepless night because of that? Bats do normally rouse themselves from time to time during the winter months, but every time they do they draw down precious fat reserves that cannot be replaced till spring. There is only so much gasoline in the tank. With white-nose syndrome, bats wake up repeatedly and fly around at times when they'd be better off snoozing. Maybe they were dying of starvation, then, brought on by an irritation that the fungus triggered. There had been experiments with fungicides applied directly onto hibernating bats—but that, too, had seemed to interfere with their sleep patterns. Many of the treated bats had subsequently died, perhaps in response to the stress of medication.

Despite this grim report, Darling managed to exude a happy-warrior optimism. He was a man in love with his job, and white-nose syndrome had made the past few years exciting. Nothing like a challenge—or a series of challenges. One of the biggest was fostering ties across the smorgasbord of government agencies—at every level—whose agendas now included trying to address the causes and/or deal with the consequences of this strange disease. Then, too, there was the community of scientists in higher education settings and at research labs. Could they all pull together? Could they share information, and coordinate their efforts? From its first discovery in Howe Caverns, near Albany, white-nose syndrome was now killing bats in nineteen states and four Canadian provinces. Each of them had its own fish and wildlife agency and its own apparatus for addressing rogue events in the ecosystems under their purview. Organizations that had never had to work together now needed to function as a well-oiled machine. And quickly, too—before the moment to do something that would help the bats had passed.

The more Darling described the bureaucratic situation, the more it sounded like a stultifying desk job. Endless phone calls and e-mails to deal with, writing up reports and reviewing those of others. Giving expert testimony before legislative bodies. Attending professional conferences and roundtables. All the red tape and turf battles that come into play when the government tackles something. It seemed obvious that Scott enjoyed the chance to get out of the office and do a bit of field work—even under bone-chilling conditions, like this morning. Sizing up the habitat potential of our forest here, and making some decisions that were likely to improve it. Again and again he'd say: "If I were a bat, what I'd do right here is—" Then he'd describe an arc with a raised arm sweeping through the air. "After that, I'd bank around this way—do you see?" We didn't, but we let him go ahead and do his thing. He was the bat man, and we watched in fascination.

When at last we got to the last pair of bat zones—the two half-acre circles that were near the farm pond, slated for enhancement as *foraging* areas rather than for roosting—Scott started marking trees with a blue X to indicate that they should stay. These were far less numerous than the trees he felt should go, so that saved time and paint. In general, he felt that any tree less than seven inches DBH—diameter at breast height—should be taken down. And that meant a lot of trees, though most of them were low-value species with crooked trunks. Unacceptable Growing Stock and culls, in forestry vocabulary. Usable as firewood, but little more than that. The goal was to leave enough canopy-layer trees to reseed the forest floor and offer some high cover, but to remove the forest's midstory trees that were presently inhibiting the bats' success at foraging. Once the work had been completed, catching bugs and gobbling them in flight would be much easier. The bats would presumably take note of their success rate, so they'd make a point of coming back here to chow down.

By the time our hike was finished, my teeth were chattering; that made it hard to speak. Scott was still gloveless and hatless, but his hands were warm. He seemed pretty stoked—*thrilled* is not too strong a word—about our forest's possibilities for bats, and he wanted to come back when the work was done and see things then. He wanted to spend another night in the woods here, trying to find out if any bats had found the roost trees. *If* the bats had found them—I took that to mean he thought there still would be some bats to find. Underneath his spokesperson's mask of gloom and doom, I could sense that he was basically an optimist. Now a passing squall of snow began to bleach the landscape, but the morning's work was done. Everybody got into their vehicles and hit the road; I pulled my chainsaw from the Batmobile's tool box and brought it in the house to tune it up and file its teeth. Starting on the first of November, we had work to do.

When you're cutting trees for any ordinary purpose, you choose ones that you know are going to fall in the direction that you want them to. If a tree's line of fall is totally ambiguous, you pass that one up. If a tree "wants" to fall into some other tree and get itself hung up there, you leave that one standing, too. Why create a widow-maker? Generally speaking, almost all the trees that I had taken down in my life had either been standing in the open or had grown up right along the forest's edge. Easy pickings, when it comes to putting them on the ground. But if you're cutting trees that somebody has marked because they interfere with sunlight shining on a shagbark hickory, it's a different story. For one thing, you'll be working in the forest's interior—where other trees will doubtless interfere with a marked tree's natural line of fall. You'll be dropping trees that do not *have* a clear fall line—or not one you can reliably predict. And you'll be dropping trees whose line of fall you damn well know is going to cause some trouble. Trees that, under normal circumstances, you'd never mess with. But I had been assigned to take down specific trees, whether they'd be easy to drop or difficult. As soon as I started on the long-awaited forest work, I realized I'd have to up my game as a logger. There were trees marked with a splotch of blue paint that seemed beyond my ability to bring down. But I had to try.

Actually, that was not strictly the case. When we had discussed the project, walking through the woods with Scott, Kate had suggested that I didn't need to actually drop every marked tree. If I felt that some were beyond my ability, I could simple girdle them; once the trees had died they would stop putting out leaves, so the goal of getting sunlight on the roost trees would have been accomplished. But when it came to each specific case, I preferred to try and put each marked tree on the ground. Girdling was not my thing, even though Brendan Weiner had advised me that it was the way to get more snags. And snags were held to be an important part of forest health. I had a somewhat different goal for my woods, though. I wanted a forest in which trees were straight and tall and *living*, not standing dead and in the process of decay. There was something else, too: I wanted the firewood. And I wouldn't get it unless trees were taken down. In round numbers, Scott had marked one hundred twenty trees for me to drop inside the bats' roost zones—and they were the priority, leaving the two half-acre foraging zones to wait until spring. Most of the marked trees were not too big to cause me trouble, and maybe half of them would be no sweat at all. So I started on those first—the low-hanging fruit. Then, as I worked up toward the bigger, more exciting trees, I'd gradually gain confidence and technical skill.

To fell a tree of any real size with a chainsaw, first you cut a V-shaped notch on the side of the trunk where you want the tree to fall. Usually this is the same side as the tree seems to want to fall, but you can fudge it by twenty-five degrees or so; that's often all you need to slip past an obstruction. The angle and the depth of the notch are adjusted for the size of the tree and how far it might be leaning. When you've got the felling notch carved the way you want it, you move to the opposite side of the tree and make a back-cut in a horizontal plane slightly higher than the bottom of the felling notch. If all goes well, the tree begins to list into the notch as the back-cut deepens. Then a moment comes when you can feel the tree begin to fall; that's when you remove the saw and quickly walk away. Sometimes a falling tree will make a dead-cat bounce when it hits the ground, and the severed trunk can fly up and bop you one—so it isn't smart to be standing in its way. As the tree falls, though, it's guided by a narrow hinge of wood that you left intact between the back-cut and the notch. The hinge is a critical part of the procedure, since it guides the tree's fall in a particular direction and also slows, to some extent, the rate of its descent. If you mistakenly cut through the hinge, you no longer are in charge of events. You can yell "Timber!" but your best bet is to run like hell. Often, a well-executed hinge snaps clean as the tree hits the ground; if it doesn't, it's not hard to saw through what is left of it.

It was thinking about that tongue of wood—the hinge—that got me to reflecting on the nature of authority. If there were no hinge, a tree could topple down anywhere. If the hinge was well-placed and not too thick or thin, a tree's fall would be effectively controlled. Put another way, the hinge exercised authority. And then, when the hinge's wooden fibers reached their breaking point, the hinge would snap and disappear. The control that it exerted was a temporary matter. It was limited and purposeful—and, one might say, benign. It was like my long runs of baling twine through the woods, demarcating what was in a bat zone and what was not. That was an expression of authority, too. But in a couple years the twine would melt into the landscape, after its work was done. What the twine was *not* was authoritarian; neither was the hinge of wood that guides a falling tree. There was a distinction here. Did I mention, early on, that my life had seemed to be a struggle with authority? Maybe that was not the case—maybe it was authoritarianism that I had been struggling with. People and institutions that had demanded I adhere to their demands blindly, and for no good reason. And at the expense of what I took to be my freedom. That was not authority *per se*, but a perversion of it. It was a big mistake to think they were equated.

Then I thought about the authority embodied by the paint splotches on these trees. It was based not on some desire to control behavior, but on a display of remarkable expertise. Uncanny expertise. Darling had walked through the woods and looked at this and that, weighing different choices and then squirting out a stream of paint. He was not trying to make certain trees behave for him, or send them *to the Foot!* in a game of "Who, Sir? Me, Sir?" He was not saying "Gotcha!" every time he marked a tree. He was just out to make the forest serve the needs of bats. He had been authoritative, not authoritarian. Same root—same root as *author*, come to think of it—but not the same idea at all. In fact, the more I thought about it, there seemed to be an inverse relationship between the two ideas. If you really were an authority on something—if you had the goods, that is—you didn't need to play that hard of a game. If you *didn't* have the goods, then playing a hard game would have to be extremely tempting; otherwise, you'd have to admit that you were flying blind. Like my father's mask of coming on like a potentate, hoping to divert attention from his insecurities. Authority—real authority—needs no disguise.

Was there an analogue in the behavior of demanding institutions? I took turns choosing one, then pointing out its hidden weakness that was kept under wraps. The military was too easy: if your mission means sending young people off to die in wars of no clear purpose, certainly you'd want to get them marching in formation and teach them to salute. And, above all else, not to question their orders. What about the vast institutions of law enforcement? If you had millions of citizens locked away—often for mere possession of a recreational intoxicant—you might feel insecure about what you were up to. You might wonder if you really were solving society's ills, or adding to them. Good reason, then, to deploy more and more armed agents to enforce the law. What about the educational establishment? Arguably, school does far more harm than good for many students. There are fewer winners than losers in most schools—and the losers graduate with issues that will haunt them for the rest of their lives. If indeed they manage to graduate at all. Understanding this, you'd think that educators would be humble. But they go the other way and claim to know what kids should do. Who should take the college prep courses, and who should not. Who should go out for a team sport, and who should not. Who should be concerned with doing well on the SAT. Who should join an after-school club to become "well rounded." Do this, don't do that. And if you reject the school's directives, those who laid them down can hurt you. When they send you *to the Foot*, the damage can be permanent.

The easy trees were coming down now, one after another. It felt wonderful to finally be logging. I could have gone on to consider other institutions that had tried to shape me—churches, for example. And the Boy Scouts. And a summer camp. But by now I was growing mentally exhausted. Bottom line: to the extent that an organization came across as authoritarian, there had to be some skeletons lurking in its closet. Something to feel bad about. Something to cover up. In a way, it answered the question that had bothered me when I weeded garlic mustard: open palm, or closed fist? As between being a gardener *in* nature and a dominator *over* nature, where would I have placed myself? Now I saw where I belonged. A gardener has the right to function with authority—*earned* authority, based on a compassionate attempt to see what's going on. A ruler—a dominion freak—sees nature as a challenge to their exercise of will. There is no compassion in their style of interaction, and they make no effort to take in the larger picture. The hallmark is refusing to ask the hard questions, or even entertain them. So I had been a gardener, crawling on the forest floor and rooting out invasives. I had been a gardener with my jar of glyphosate, too. And now I was gardening with a chainsaw, felling trees that shaded certain other trees where bats might want to roost. It was a gesture of authority, sure—but it was not authoritarian. What I was doing was limited and purposeful. And, in a way, benign. I had nothing that I needed to feel bad about.

Even under the best conditions, work like cutting trees is never really safe. The best thing I could do to protect myself from harm, I figured, was to take it slow and easy. And to stop when I felt tired, when my muscles told me I'd begun to lose my edge. So after the low-hanging fruit had all been harvested, I'd never drop more than two or three trees per day—early in the morning, when my mind was alert and my body felt fresh. Then I'd spend the balance of the session cutting those trees up and dealing with the pieces. That sort of work was a good deal safer, and at times it was essentially mechanical. Depending on the species, there might be a hundred branches to cut into stackable lengths and pile up. Then there were those portions of the tree that had some fuel value—by my calculus, anything thicker than about two inches. Smaller sticks could serve as kindling, and a good-sized slice of trunk would hold a fire overnight. Sometimes the lower sections of a tree were thicker than would fit into our woodstove, so I'd have to split those into several smaller pieces. Time and distance, over and over again. And there was no serious pressure to hurry up; I had until April first to open up the roost trees. After that, bats would start returning to the woods.

Day by day, I felt I was improving my confidence and working with increasing skill. Sometimes I could drop a tree with breathtaking precision. But then, just a couple days before Thanksgiving, I misjudged a white oak and it sat back on the saw. The tree in question was fifteen inches DBH and forty feet tall; I had figured that it wanted to fall westward, but it turned out I was wrong. Once I had the saw's blade buried in the back-cut—coming in from the east side of the tree—the trunk revealed its actual center of gravity by settling its considerable weight onto the saw. The chain stopped moving, and the blade became pinched beneath half a ton of hardwood. First I swore; then I reached to switch the engine off. This was the first time I had made a serious mistake. And it was a bad one, since I no longer had control over the tree.

The first priority in such a situation—once your heart stops pounding, and you start to catch your breath—is to somehow get your saw out. One way to do this is to drive a narrow wedge into the kerf the blade now occupies. If you can widen that crack by a quarter inch, the bar and chain should slip right out and be none the worse for wear. Thin, narrow wedges to accomplish this are made out of super-hard plastic, since using a steel wedge could damage your machine. I used to have some plastic wedges, in my younger days—but I had no real hope of finding them now. And a trip to the nearest store where I could buy some would take at least an hour, during which time the tree might topple on my saw. That would likely be expensive. So I drove the Batmobile home and poked around for something I could use to improvise. What I came up with was a pair of hardwood wedges used for fastening a sledge hammer's handle to its head. They weren't much to work with, but I thought it worth a shot. Sure enough, I got those wedges driven in the back-cut's kerf just enough to liberate the saw. Then I called it quits for the day. I vowed not to come back to that part of the woods till I'd seen the oak had fallen.

Next day, I found that it had toppled of its own accord. So I cut it up, and after that I had a talk with Cheryl. "I've taken down all the easy trees," I told her. "Now I'm dropping hard ones, and it's scary. I could use some backup."

"You think I can back you up?"

"Just having you there would help. If a tree starts falling in the wrong direction, you could yell. And if something bad happened, you could get me out of there."

"How?"

"On the ATV."

She laughed. "You haven't shown me how to drive it yet."

"Well, you could bring your phone and dial 9-1-1."

"Maybe it's time for you to get your own telephone."

She had a point; I'd reached my comfort level with communications technology ten or twelve years ago, and since then I had not kept up. Now I seemed to be the only person on the planet who did not carry a cell phone. "But if I'm knocked out," I said, "a phone isn't going to help."

"This is that dangerous?"

I nodded. "There's like thirty trees left, and every one's a challenge." There were trees marked for removal that were bigger than anything I'd ever taken down. In my life. There were trees with dead limbs that might break off, drop fifty feet and conk me on the head. There were trees perched on steep slopes with rocky soils—places where the ground wasn't stable underfoot. So it wasn't smart for me to tackle them alone.

"If it's all that dangerous, maybe you should hire an expert."

Ouch, I thought. That hurt. "Maybe I should," I said. "But this is my project. I would like to see it through. And I'm pretty sure that I can do it—I just want you there."

So next day, Cheryl rode out with me on the Batmobile. I dropped a pair of maples that were clearly dominant, reaching up fifty feet or more and commanding their share of forest canopy. Each of them would take a couple days to clean up. And that was the pattern as we moved into December: drop a couple of marked trees with Cheryl there watching me, then drive her back to the house and return to start reducing them to firewood. By now I was working in heavy winter clothes, but apart from the bite in the air it didn't feel like winter. There had only been

one serious snowstorm, and the stuff had melted quickly. As long as the ground was bare, I'd keep going to the woods—though I'd come back freezing after just a couple hours. Still, two hours a day amounted to something. I was making steady progress. If I stayed on that pace, certainly I'd have the roost zones ready by spring.

Trying to stay mentally active as I piled wood, I kept returning to my thoughts about authority. And now it was time to raise the issue in my father's case, since he'd been the catalyst for much of my rebellion. My father had legitimate, earned authority in many areas of life. He had a thorough understanding of electric circuits, from those used in heavy machinery to those hidden inside electronic gadgets. He could take apart and fix any home appliance. He knew how to cook and sew. He knew how to garden. He could write computer code. He could design and build things out of wood and brick, concrete and glass and steel. Given the range of his areas of expertise—the fields in which he *was* an actual authority—there was no reason why he had to adopt an authoritarian persona. A man who, if questioned as to why somebody had to do exactly what he'd told them to, would snap: "Because I said so!" That was a tragic flaw. I could understand, now, that his well-concealed insecurities were to blame: a lack of inner confidence based on certain things that had happened to him as a child. His father's early death, for starters. Then his mother's inability to care for him, and her ongoing battle with depression. His stepfather's insistence that he change his last name and thus deny a crucial part of his identity. Still, it was pathetic that a man of my dad's competence—of such a range of competencies—could have become so bent on telling people what to do. And then making his demands a test of fealty.

There came a moment when I'd finally had enough, and to some extent what happened next altered things between us. It must have been at least twenty-five years ago; my parents had been visiting the farm, and we'd gone out to dinner. In the special glow he got from picking up a restaurant check, my father had turned feisty as I drove us all home. Maybe he had had a glass of wine with his meal, too—that might have set him off. Anyway, he wanted me to know that the water pump in our basement was making a funny noise. Making a racket when

it first started up. It shouldn't sound like that, and it needed looking at. The implication was that I was not on top of things. Then there was the matter of some loose shingles on the roof. Wasn't I aware that if I let that go, I'd have a leak? Once that started, things would quickly go downhill. I endured the growing to-do list he was laying out, biting my lip as I had learned to do. But then as I turned the car into our driveway, it jounced over one or two depressions where the gravel had become compacted. Not exactly potholes, but places where the surface could have used some topping up. "I don't know why you don't fix these ruts," he told me in a certain tone of voice—whiny and sarcastic. Also spoiling for a fight.

Right then, I stopped the car. "Dad," I said. "I'm sure that you could live my life a whole lot better than I can. But the thing is, you don't get to."

That shut him up. Next morning, he was out to win me over: making a pot of coffee before I was out of bed, building a stair railing, engineering modifications to a farm machine. It would not be like him to apologize—not to me—but his life had taught him that affection, once squandered, could be earned back by doing nice things for those he'd wounded. Making them grateful for his efforts in their behalf. That was his *modus operandi*, I now realized; first he'd draw you near to him, then he'd say or do something to drive you away. Then he'd set out to prove that he could win you back again. It was a loop that he was stuck in, an endless cycle. And I realized that I had often done the same thing. I could think of girlfriends whom I'd treated abominably, at the same time as I was swearing that I loved them. Then I'd work to get myself back in their good graces. I had done it with certain family members, too, and friends; certainly I'd done it with Cheryl and our children. It was a way of responding to other people that I must have learned at my father's knee: win affection, then betray it, then win it back again. What, after all, are children studying when they watch their parents interact? How to win another person's love—and then, having won it, how to stay empowered in the relationship. For my father, mistreating loved ones was empowering. If they then rejected him, he would work to win them back—and then be re-empowered. And then do it all again.

By mid-January of 2012, it was becoming clear that winter wasn't going to happen this time around. Not in Vermont, at least, or even in the whole Northeast. It had gone on holiday—to Russia, based on news reports, and parts of Eastern Europe. After the previous year's run of savage weather, we were getting off without any snow at all. Road crews had enormous piles of sand and salt socked away. Keeping our driveway open this year was no problem. There were occasional cold nights, but nothing that would make us break out extra blankets. Most days, I could do my two hours in the woods; the work was going smoothly, and the cords kept piling up. Around the shagbark hickories there was noticeably more space and light. When the bats emerged from their caves in the spring, hopefully they'd find our opened roost trees and move right in.

Then one morning—it was on January 18—a front-page newspaper story put the nationwide death toll from white-nose syndrome at between six and seven million bats. That was the U.S. Fish and Wildlife Service's brand new estimate, and it was a heck of a bump from the previous figure of one million. Right away, spokesmen for the spelunking community challenged the figures as deliberately alarmist; they wanted to get back into caves that had been placed off-limits. That did not seem likely, though, anytime soon. In Vermont, three more bat species were placed on the state's endangered list, bringing the total to five species of out nine. Even if a cure could be discovered right away, it might take a couple hundred years for the little brown bat to recoup its former numbers. By then, who could say what other changes would have taken place? Maybe unaffected bat species would expand to claim the empty niche, but no one knew for sure. Two hundred years amounts to several human generations. And it's more than half the time since Pilgrims stepped on Plymouth Rock.

As to how the G. *destructans* fungus had arrived here, current speculation was that a European bat might have come over in a shipping container—as a stowaway—and then escaped into the wild when the container was unloaded. A single bat might have set the wheels in motion for what was now an environmental train wreck. Then, too, there were new theories as to why the fungus wasn't killing bats in

Europe. Maybe they had coevolved there, millions of years ago. Maybe North American bats had evolved in the *absence* of the fungus, and thus had no resistance to it. None whatsoever. No reason, in other words, to even have a cohort of their population that might ultimately prove resistant. If that were the case, things might get very bad indeed. At Carlsbad Caverns in New Mexico—two thousand miles away—visitors were now being screened and given plastic booties before being allowed to venture underground. So far the bats there had not shown signs of white-nose syndrome, but there were suggestive signs that it had reached Oklahoma. Not that far away. In the minds of some concerned biologists, white-nose syndrome had already become one of the worst wildlife health crises in recorded history. And it was still only five or six years old.

At some point in the winter, I asked Kate Teale to come out and certify the work that I'd been doing. The daylighting of roost trees had been scheduled to unfold over two separate winters, but the first half of the job was done and I had a check coming. If my work passed muster, I planned to start on the second half right away—including the foraging areas, the all-night diners to be carved out of the woods adjacent to the pond. With luck, I could get the whole job finished before spring. So Kate and I took another walk through the woods together—this time, bundled up—and the first thing she told me was that Toby Alexander was surprised at how few trees Scott Darling had marked for removal. They didn't seem that few to *me* at this point, but I got her drift. And one of its corollaries was that if the people who designed this project had known how few trees would be slated for removal, maybe I would not have had to pull out all that garlic mustard. Maybe my war against invasives had been waged on a wider scale than had strictly been necessary. Maybe I had weeded several acres of the woods for nothing.

That was now water under the bridge, though. And I let it go. The bat zones I had opened up were looking terrific, apart from three or four marked trees that I had overlooked. But they weren't big ones, and I promised I would have them on the ground before the day was done. Kate took GPS coordinates for each of the twenty-nine shagbarks I had thus far daylighted, and she snapped photographs of various stumps with their splotches of blue paint. Evidence that the slated work had been accomplished. Then she asked me what I had done with all that wood.

"There must be three cords piled by my garage," I told her. "And five more stacked up in the hay barn."

She nodded, but I could see she wasn't cool with this. "I'd have thought you'd leave more trees lying on the ground."

"Why?"

"We like to see at least four down trees per acre."

"Why?"

"Because it's beneficial. Down trees are good for the forest. For its general health."

"W-w-w-wait," I stammered. "B-b-b-but—" And then for a couple long seconds, my voice froze up. I realized, of course, that I had just revealed my worst fear. I had stuttered badly—twice—and now Kate was bound to regard me as speech-impaired, with all that implied about my personality. But I could not help but stutter, under the circumstances. After all the work I'd done, had I broken some new rule by using trees for firewood? That had not been spelled out in my contract with the folks from WHIP. And I knew what Brendan Weiner had put in our Forest Plan about having "down" trees: he had wanted *two* of them per acre, and we had that many. What my forest lacked was snags—which fall apart over time and thus keep the supply of down trees replenished. And he had told me I could get more snags by girdling. I took a deep breath, getting a grip on things. "Sounds like you wish I'd girdled more trees," I said to Kate. "Rather than just felling them." That was disingenuous; I hadn't girdled any.

"That might be a good idea, looking down the road," she told me. She wasn't going to be a hard-ass over this. Maybe in the work that I had yet to do, I'd leave more wood to rot on the forest floor. Down trees played a role in nutrient recycling and soil formation, not to mention offering food and shelter for a host of organisms. Basically, though, I'd done an adequate job and I deserved to get my check: five hundred eighty more dollars from the taxpayers. Oh, and there was one more thing. Daylighting the roost trees was an exercise in habitat enhancement, obviously. But before I went ahead with carving out the two bat *foraging* zones—which would open up that portion of the woods dramatically—I needed to describe the planned work to Chris Olson at the county forester's office. He'd need to sign off on that work before I started.

"Why?"

"Because it isn't a management recommendation in your Forest Plan."

"The Plan calls for conserving and enhancing habitat for bats. Doesn't habitat mean food, as well as shelter?"

"It would just be smart to let Chris know about it, first." Though Olson was working for the state rather than the feds, he did have some jurisdiction. If I reshaped the woods in ways that hadn't been specifically laid out in our Forest Management Plan, there could be repercussions. Technically, it could be construed as a breach of contract—and that would jeopardize our standing in the state's Use Value Appraisal program of tax relief. Kate reiterated that since providing areas for bats to fill their stomachs in—as opposed to finding roosts—had not been explicitly stated in our Forest Plan, it was important to get Chris Olson's approval.

After Kate had driven off, I had a mini-meltdown. Who the hell's forest was it? Did I now have to get the government's permission before cutting my own firewood, from my own trees—which I had taken down per the government's directives? That was preposterous. And since our Forest Management Plan had made "bat habitat" our primary mission, why didn't that give me rein to carve out feeding grounds? Why should that require preapproval by some functionary working for the state? No wonder I had stuttered, and then felt my voice seize up. It was the injustice of it. How could I be asked to do my job of fixing up the woods and also stay on track with all these branches of the government? It was—well, it was batshit. How many people in a million would devote this time and energy to helping bats—and then be warned about not doing it in a bureaucratically acceptable way?

But yes, I had stuttered—and having done so gave me pause. The idea that I'd entertained a few months earlier was that my speech impediment had come on at a tender age by playing "This Little Inch" with my Gramps. Maybe that was not the case, though. True, as a child I had had a nagging stutter; true, I had been fondled by him. But if I could stutter as a grown man over the government's questioning my right to cut some firewood—well, then. Maybe my childhood stutter had some other cause. Maybe I'd been set off by something that had struck me as unjust; I *am* overly sensitive that way. I did not forgive my Gramps for the things he'd done to me, but there was no reason to believe he'd caused my stutter.

Now something else that had been bugging me unconsciously emerged and could be dealt with in a straightforward way. One thing that had seemed outrageously unjust in my childhood memory—recently brought to light—was that I'd been left in the care of a pedophile while my parents took my older sister to a football game. If my father had a clue—as Anaïs said he must have had—what kind of games my Gramps liked to play with little boys, then it had been criminal to leave me alone with him. How could that have happened, if my father knew what Gramps might do? But then it occurred to me that my dad was ten years old when his mother made her second marriage with this troubled man. Ten years old, not four. My dad had never interacted with him as a four-year-old. Even if my Gramps *had* tried to grope my father, he might not have thought the man would try it with a four-year-old. That was something that I didn't know, and had no way to prove.

Then a last thought about my father came to mind. Earlier, I said that he'd uprooted our family from a vibrant web of kinship in Chicago, in the 1950s; moving to New Jersey was perceived as a betrayal, and I blamed it on his trying to challenge his children with a life-shock like his father's death had been for him—and at about the same age. But maybe it had been something different than that. Maybe he was out to spare us from the role that Gramps would want to play in our lives—or at least would try to play—when we had turned ten or twelve. When we were the age at which my father came to know him. Well, people rarely make a life change for a single reason; I was well aware of that. But something was terribly "off" about my Gramps, and perhaps my father wanted to protect us from his influence. One way to do that was to move his family far away. Could I know that that was true? No, but it was possible. There are times in life when being generous about someone's motives costs one nothing, and this was such a time. Had I had a better chance of turning out the way I did—and turning into *who* I did—by being removed from the orbit of Gramps's influence? Yes, undoubtedly. And had my father made that happen, by moving us? Yes, he did. Absolutely.

As the non-winter wore on, I did get in touch with Chris Olson at the county forester's office. I said I'd been told I had to get his permission to create two areas where bats could hunt for insects, as per the contract I had entered into with the Wildlife Habitat Incentive Program. I told him that Scott Darling had been shown these potential bat foraging zones; he had agreed they would be useful to the bats, and he had told me which trees I ought to cut. Amazingly—or maybe not amazingly at all—Chris Olson's office had not been alerted to the complicated project I had undertaken with the feds. He was just a state employee, after all. He had approved our bat-based Forest Management Plan in the year when Brendan Weiner first presented it, but he didn't know what I'd been up to since then. All I had to do, he said, was bring him a map of where the work was to be done and a copy of the project's official description. Which I did, and happily. One more agency of government satisfied.

But then a few days later, Chris called me at home and said he wanted to come out and make a Field Inspection; this was a drill he sometimes had to put folks through. So I took another walk through the forest with a person in the taxpayers' employ; by now, I had a pretty good rap to lay on people as I led them around. Olson was hatless and gloveless on a frigid day, and he exuded a no-nonsense attitude. Either I was following my Forest Management Plan, or I was not; that was what he'd come to see. And long before the tour was over, he told me that he felt I was. He was particularly pleased that I'd had Scott Darling come to mark specific trees, and he was impressed by how I'd pushed back against the buckthorn. Garlic mustard didn't seem to bother him at all. But he had a woodlot of his own where he said that glossy buckthorn was thriving, and he hadn't whacked it down yet. He hadn't found the time. Then he told me buckthorn had been actually *promoted* by some government agency, in the not-so-distant past. Ditto for multiflora rose, another bad invasive. That one had been touted as a means of growing "living hedges," till it ran amok and started taking over pastures. These noxious plants hadn't surreptitiously invaded; rather, they'd been introduced. For landscaping purposes. That threw me for a loop—one generation of government programs setting bear traps for the next.

A few days later, a "Conformance Inspection Report" came in the mail and our forested acres were recommended for continuation in the Use Value Appraisal program. That was a big relief. But I was supposed to begin filing annual Management Activity Reports with Chris Olson's office. Otherwise, how was the state to keep tabs on me? Now that all my government ducks were in a row again, I proceeded to start carving out the two bat foraging zones. Lots and lots of trees came down—at least a couple hundred, though nothing with a DBH of greater than seven inches. Here, though, the woods *were* really being opened up; here the tight control over invasives was important. When the work was finished, dominant trees would be pretty much all that was left—plus a few snags and some promising saplings. And I'd have another five cords or so of firewood.

At the end of February, Anaïs's new CD came out with my picture on the cover. My iconic photograph as an angry young man, to illustrate her title song "Young Man in America." There was an insert with the printed words to all the songs, songwriting being among other things—an act of poetry. And as I had reason to suspect, several lyrics dealt with my interactions with my father as seen through her eyes. And she got a lot of things right; in many ways, our difficult relationship was newly illuminated. The song "He Did" struck me as especially potent, ending in a litany of questions aimed straight at me:

Who gave you an ax to grind?
Who gave you a path to find?
Who gave you a row to hoe?
Who gave you your sorrow?
Who gave you the break of dawn?—
(a pleasure just to look upon).
Who gave you a barn to build?
And an empty page...to fill.

What she means, I think, is that each person should be grateful to those who gave them something to chafe against. Even if they can't say thanks. And that it is ludicrous to rail against your father because if it weren't for him, you wouldn't even be here. Wouldn't be around to feel angry, or rebellious, or joyful—or anything. Wouldn't even *be*, in fact. Which is true. I understand that. And yet people are responsible for what they do. If we want a less authoritarian world, it will have to start with *people* learning to be less that way; I can't believe it's going to start with institutions. *People* need to learn to stop behaving like the government. And when people do behave like the government, I don't think it's wrong to call them out on what they're doing. Understand where they're coming from, sure—but at the same time keep saying *No* to their demands. That's something that I'm glad my own kids have done for me, when I well deserved it. And I likely will again. My father had a lot of traits I couldn't help inheriting—many of them admirable traits that have served me well. But there are some others I'd prefer not to pass along, though I realize it's probably too late for that. With luck, perhaps my grandchildren—or my great-grandchildren—will grow up unaffected.

Walking in the woods the other day, I felt a scary freedom—scary, and yet exhilarating too. I now knew them intimately, inch by inch; getting to know them, I had stared hard at myself as well. Now the work was almost done. And would my efforts help the forest save endangered bats? It almost seemed beside the point. Bats might find the shagbark hickories and I'd never get to see one—never have the certain knowledge that they had made roosts there. Or the bats could be wiped out and scarcely leave a trace. Buckthorn might come back with a vengeance and reclaim its space; garlic mustard could quite easily crop up again. What would stay with me were my echolocations, though: a sixth sense that had helped me navigate the skies of psyche. And the sense that things do change—that everything is changing, and so people may as well change, too. Why, given time enough a hand can grow into a wing. And when it does, it's going to find a way to fly.

Field Notes

On June 11, 2012, Scott Darling called to say he planned to send a colleague to the farm soon—perhaps that evening—to place a "bat detector" somewhere near the pond. We were in the middle of bringing in hay, and I was exhausted but I stayed awake till nightfall in hopes of meeting up with this wildlife technician. Her name, Scott told me, was Alyssa Bennett. I never saw her arrive or depart, but next morning there was a contraption sitting on the dam: a black box propped up on an easel made of two-by-fours. Its job was to record the high-frequency cries that bats emit while hunting insects. Since the cries of each bat species have distinctive acoustic features—including frequency, intensity, duration, and pulse interval—once they are recorded they become a kind of signature. After reviewing a couple nights' data on the bat detector, Scott said that a team would be arriving around dinnertime on Thursday, June 14, to set up mist nets. He himself planned to show up around ten o'clock—once the bats would have left their roosts and started prowling—to oversee the trapping. He said I was welcome to observe the proceedings, but that I'd be smart to get a nap in. His crew stayed up late.

Five people came on Thursday evening to erect the nets, converging here from five different corners of the state. Their leader seemed to be Alyssa, who was a remarkably slight and slender woman. Just a wisp, really—but a highly energetic one. She had scoped the premises and made certain choices as to where the different nets should go. But Joel Flewelling, another of the state's team of wildlife technicians, seemed to be in charge of physical assembly of the complicated apparatus that held up each net. The nets themselves were woven out of some black gossamer, arrayed in a chicken-wire pattern that the bats could not detect until they were ensnared. The biggest of these nets, which measured thirty feet wide by thirty feet high, required all six of us to get set up and safely braced. That's right, six of us—I was tapped to help out, too. The net was made to stretch between tall metal poles that were guyed to several ground stakes, and a system of something like vertical curtain-hangers enabled the net to be lowered to chest height when and if a bat became caught in its upper reaches. Otherwise there would be no way to get it out.

Some of the members of our team were clearly interns, almost as new to this business as I was; others had spent many long nights trapping bats—or at least trying to. In the current season thus far, the pickings had been slim. Joel said that last time out, they'd only caught a flying squirrel and it did considerable damage to a net. But they seemed to have high hopes for the night ahead. By nine p.m., we had four mist nets rigged and ready. The biggest of them stood right at the entrance to one of my foraging zones, where I had cleared away midstory trees to make it easier for bats to run down bugs. The other three nets were in the woods along the VAST trail used by winter snowmobilers.

With the nets in place, we gathered at the Fish and Wildlife truck—parked atop the pond's dam—and swatted at mosquitoes for a good half hour. We compared the merits of different brands of bug dope, but it was conceded that bat research meant giving blood. Where the air is thickest with mosquitoes, that's where bats will feed. The bed of the truck was crammed with sophisticated instruments and field gear; it was a rolling biological laboratory. Headlamps were dug out and duly distributed, and I was shown how to use mine by a woman named Zapata. She spelled out her name when I asked her to—yes, her namesake was the Mexican revolutionary hero. She was the same age as my daughter, roughly; I figured that her parents must have come to Vermont as hippies in the 1970s, too. How else could she have been saddled with a name like that?

Just as dusk was settling in, Joel went to check the nets. It wasn't fully dark, but he shouted "Bat!" as soon as he reached the net that stood astride my foraging zone. We all ran to see; sure enough, a ball of fur was nested in the web of fabric twenty feet above our heads. The net was gently lowered, and Alyssa—wearing latex gloves—grasped the bat delicately and used the point of a sharpened pencil to release it from the strands of netting it was tangled in. That done, she put it in a nylon mesh "bat bag" and hung it from a rack over the bed of the pickup truck; that gave the detainee a temporary roost from which it could not escape. As night made its full arrival, one more vehicle pulled into the driveway. Soon a flashlight beam came weaving down the hill toward us. Scott Darling greeted his colleagues from the darkness, then emerged into the pool of light shining on the truck bed. He saw the bat bag right away, and smiled. "Looks like we're in business, huh?"

"It's a big brown," said Joel—as if this were nothing special. "It's a male, too."

"Let's see."

Darling put on latex gloves, then reached into the bag and carefully removed the bat. Maybe it was big compared to little brown bats, but with its wings folded tight against its torso the creature was no bigger than the top joint of my thumb. It was a homuncular Icarus, I realized. Complete in all details and anatomically correct. "Look," I couldn't keep from pointing out. "Is that a tiny penis?"

"No," Alyssa told me with a laugh. "That's his *giant* penis."

Darling said that big brown bats, for some undetermined reason, had not been decimated by the white-nose syndrome to nearly the extent that several other species had. Finding one was good news—finding *any* healthy bat was certainly good news—but not quite cause for celebration. The bat started making little *chuh-chuh-chucking* sounds, out of either rage or fear; then it tried to bite right through the glove that was holding it. Scott blew on its furry head to chill it out, to calm it down. Now the task at hand was to "record" the bat, which meant weighing it on an electronic gram scale and peering through the membrane of the wings to check for signs of damage. Healthy wings were assigned a score of 0, and a badly war-torn wing might be rated 2 or 3. Then a tiny metal band was placed around the forearm of the bat, allowing room to slide back and forth but not fall off. The relevant details were entered in a field notebook: *Eptesicus fuscus. Male. 14.6 grams. Wing score = 0. Band #43206. Caught at Mitchell's, 6/14/12.* Then the bat was set free to go about its business; after perching tentatively in the palm of Joel's hand, it flew off into the dark.

As a protocol of humane wildlife research, the nets had to be checked at fifteen-minute intervals to make sure that no bats would be caught in them for long. And for the next hour, nearly every trip to check the nets produced another bat. Five of them, in all—and each another big brown bat. Two were pregnant females, based on their weights and on the ripeness of their bellies. One managed to bite its way through Scott's latex glove and draw a tiny drop of blood; that one had to be returned to its bag and observed for behavioral signs suggesting rabies. Roughly one percent of bats are thought to be rabid, so anyone who handles bats has to have been vaccinated. Scott was up to date with his shots—but even so, he said that rabies was something that he didn't want to mess with. If the bat did act like it was rabid, it would have to be destroyed and then tested. If the test proved positive, he'd have to see a doctor.

But the night was going well, as far as big browns were concerned. And then, around ten-thirty, Joel cried "Bat!" from up in the woods and returned to the truck with something that was *not* a big brown bat. Darling seemed to recognize the species right away, but he let his colleague work through a decision tree used to arrive at a positive ID. At first glance, Joel thought the bat might be a little brown—and given their diminished numbers, that would have been a find. But did it have a "buffalo" head, as though someone had recently punched it in the nose? Yes, in fact it did. And did it have pinkish lips? Yes—that, too. And did the brownish toe hairs fail to reach beyond the toes? Yes—they were considerably shorter than that. And was there a keel—like the bow of a canoe—to the shape of the calcar? Yes, indeed there was a keel. Though the entire calcar was the merest nail clipping.

"So what do you think?" asked Darling.

"I'd have to say this is an Indiana bat."

"I would have to say that, too."

Everyone gathered around to share the moment, and everyone—in his or her own way—expressed their jubilation. Indiana bats had been declared endangered back in 1967, long before the ravages of white-nose syndrome. Now they were as rare as hens' teeth. Here we were, though, staring at one in the light of several headlamps. "I love you," said Zapata to the startled-looking bat. And it was a female, capable of reproduction. Then Alyssa weighed it in at 8.3 grams; based on that weight and the distension of its belly, the bat was likely pregnant. That made it a candidate for getting outfitted with a radio transmitter for telemetric tracking—and that, if the procedure worked, would make the night a home run for everyone concerned.

Joel and Alyssa both sprang into action. The radio transmitters cost one hundred fifty dollars, and there's no way to get one back when its time runs out. So they have to be employed judiciously, sparingly. Tiny and ultra-lightweight as they are, the battery is not engaged until it's time to actually put one into service—and to do that, two fine-gauge wires needed to be fused together. In the dark, on the tailgate of a pickup truck. The transmitter was clamped into the sort of vise that I'd seen anglers use for tying flies, and a propane torch was fired up. That was Joel's job, as well as tuning a receiver to the frequency at which the device would send out steady beeps; meanwhile, Alyssa was clipping a patch of fur from the bat's hairy back with surgical scissors. The hairs were carefully set aside for safekeeping. Then a blob of special glue was placed on the shaved spot, and the transmitter was affixed. It terminated in a slender translucent wand—seven inches long, at least—from which the signal would be broadcast for up to a mile. Then the shaved hairs were glued back over the transmitter.

All this effort took the better part of an hour, and the bat was starting to show obvious signs of stress; meanwhile, the night air had begun to grow chilly. "Bat work" has to stop when the ambient temperature drops below fifty-four degrees Fahrenheit, and the digital thermometer hanging on the tailgate was reading fifty-four and six-tenths degrees. The bat grew calm—so calm, in fact, that it was cause for concern. Alyssa likened its predicament to alien abduction; nothing in its life could have prepared it for a night like this. But the work was going as smoothly and as quickly as she felt it could. When it was time to finally release the bat, she set it in the open palm of her hand—and on take-off, it plummeted straight to the ground. That sent everyone into a panic. But Joel found it floundering in the mown grass and took it in the cab of the truck to warm it up. Then he fed it several drops of water from a tiny straw. An impromptu conference was held about what to do; the upshot was that Darling would take the bat home for the night—home to Cuttingsville, at least an hour's drive away—and do what he could to get it back on its game. Next night, if all went well the bat would be brought back here and released where it was captured. Then we'd all find out if she'd survive her ordeal.

By two a.m., the set was struck and all the gear was packed onto the state Fish and Wildlife truck; everybody got into their vehicles and drove away. I walked up to our house awash in night sounds: barking frogs, an owl's soft *whoo*, the moans of hungry coyotes lurking in the distance. It's another world at night. And I had to recognize that bats were out there, too. I mean, they *were* there—I had finally seen them. Stared into their beady eyes. That was something wonderful. I couldn't keep from smiling. Bats were on our farm and in our woods, doing their ancient thing. I had not confessed my love directly like Zapata did, but I felt it just the same. I was now in love with bats.

The next night, Alyssa showed up with a nylon bat bag just as dusk was settling in. Our Indiana bat was in it, and she seemed to have recovered. Released in the woods at the same spot where she'd been captured, the bat flew off and disappeared—hungry, no doubt, and eager to start catching bugs. Outside the house, Alyssa turned on a receiver and unfolded the arms of an aluminum antenna. A steady beeping came from the direction of the woods, which meant that the bat's tiny radio was transmitting. The battery was apt to work for two weeks before going dead; for two weeks, then, I could expect someone to show up every evening to go find the bat. To find, that is, the tree that she had chosen for her roost that day. You did this by lugging the receiver and antenna to the source of the bat's transmission, guided by its signal strength. And then, when you found that tree, you waited there till nightfall to conduct an exit count.

"A what?" I asked Alyssa. I was listening, but she'd lost me.

"At this stage of pregnancy, she'll choose a different tree each day. But then she'll start to find some other Indianas to hang out with—so there should be several bats emerging from her tree at night. We show up and try to count them, till it gets too dark to see. The exit counts should go up as she gets close to giving birth. When she does, she'll be in a maternity colony. They'll all choose one tree to raise their pups in—like a shagbark hickory. If we can track her to that tree, we've hit the jackpot."

"Wow," I told her. "This is like the coolest thing ever."

Then Alyssa had a wise and generous idea. Rather than having to hand off the receiver from one researcher to another as the days went by, she would leave it sitting on the workbench in our garage. She showed me how to turn it on and sweep with the antenna till the *beeps* came loud and clear. Then Cheryl and I could check the bat's general location every morning—after breakfast, say—and report our findings to Alyssa via e-mail. That was amazing: I could stand in my driveway with a cup of morning coffee and track this rare, endangered bat roosting in our woods. In one of our own trees.

A couple times, I tagged along with whoever showed up to make the evening exit count. Once it was with Scott Darling, who tracked the bat's signal to a long-dead elm. A snag tree. That was as close as telemetry could take you: you could stand at the base of a tree from which the signal was clearly emerging, but you wouldn't see the bat until it left its roost to start hunting. You'd catch just a glimpse of it darting off into the sky; then the beeping signal would begin to travel far and wide. What was disturbing to me, though—amid my high excitement—was that our bat was never choosing trees that I'd released. Never choosing shagbark hickories in my designated zones. Not to worry, Darling told me. After all, we'd only strapped a radio on this one bat. Many more were likely out there, each making decisions about trees that offered daytime roosts. Habitat was habitat, and I had enhanced my share. Sooner or later, the bats would find my shagbarks.

On the evening of June 20, Alyssa made the exit count and noted four other bats emerging from the same tree that ours had chosen. So perhaps the pregnant ladies had begun to congregate. But next day a nasty heat wave moved in, pushing daytime temperatures well up in the nineties. At ninety-six degrees, the glue that holds the radio device to a bat's back can soften to the point of melting; this can cause a bat to drop its transmitter, and if it does—well, Game Over. That's what apparently happened on June 22: night fell, but no bats flew out of the tree from which the signal was emerging. Same thing the following night, and on the next. On June 26, Zapata stopped by to pick up the receiver and antenna. They were needed somewhere else. For the time being, at least, we were done with tracking bats.

But there had been bats to track—that much was now established. And perhaps my efforts in the woods had helped them with their plight. In any case, I hadn't hurt them. If there turned out to be a happy resolution to the loss of seven million bats—if they turned the corner and began to bounce back—well, I had been part of that solution in a quiet way. I had been their gardener, their nurturer. And their friend. Thinking over what I had accomplished in my life thus far—if what one person can accomplish even matters—that was no small achievement. Actually, more than that. Befriending bats had been a means to figure out, against all odds, where in the world I actually was. And exactly who I was. And to participate—thankfully, joyfully—in the wild party that keeps going on around us. What I mean, perhaps, is that I'd learned to fly less blindly.

Acknowledgments

Based on Scott Darling's review of his field notes from the summer of 2006, I can now state with confidence that it was my daughter-in-law, Susannah McCandless, who first encouraged him to trap bats on our farm in June of that year. Had she not extended that invitation, the present book would not exist.

It was Llewellyn Howland, my former editor and longtime friend, who suggested to me in 2011 that the "bat project" I was engaged in here might have the makings of a book. That thought had not previously occurred to me—but once he planted the seed in mind, it took root quickly and flourished.

The book also would not have existed without the wise counsel of David Brynn and Brendan Weiner of Vermont Family Forests, who encouraged me to pursue a Forest Management Plan for our farm that would foreground wildlife habitat—and bats, in particular. Then Brendan placed me in contact with various government agents who were eager to support that work and helpful in doing so. David, too, read the book in draft and vetted its discussion of forestry terms and practices.

Besides being Vermont's chiropterist *extraordinare*, Scott Darling took time to read and review the book's discussion of all things pertaining to bats and the history of white-nose syndrome. If Vermont's endangered bat species do stage a comeback, he and his colleagues at the state's Department of Fish and Wildlife will deserve enormous credit.

I have been fortunate to have many thoughtful readers of the work-in-progress, many of whom made helpful suggestions and/or asked questions that guided my revision process. In particular, I'd like to thank John Elder, Arthur Johnson, Susan Scott, Reid Kempe, Shel Sax, and Fran Putnam for their detailed commentary.

Sally Brady undertook to represent this work to publishers without ever having met me, and she had the wisdom to show it to Margo Baldwin at Chelsea Green. I knew that the book had found the right home as soon as I walked in the door to this publisher's suite of offices. And that, when Brianne Goodspeed took charge of the manuscript, my words would be in good editorial hands.

Finally, thanks to my dear wife, Cheryl, and to our marvelous son and daughter. I've actually tried to model my prose style in this book on that of Ethan's, and I hope that I've captured some measure of his delightful wit and wide-ranging, fearless intelligence. Special thanks to Anaïs for allowing me to quote from the lyrics to two of her songs, and to consider their implications. Long journey, loved ones! And perhaps the best is yet to come.

About the Author

Author photograph by Ethan Mitchell

Don Mitchell is a novelist, essayist, and sometime screenwriter whose most recent books are *The Nature Notebooks* (a novel) and a guidebook to Vermont in the Fodor's Compass American series. He's also the architect and builder of over a dozen low-cost, energy-efficient structures on Treleven Farm and a shepherd with thirty-five years' experience managing a flock of sheep there. One of his current interests is forest management with the goal of enhancing habitat for endangered bats.

From 1984 to 2009 Don taught courses at Middlebury College, primarily in creative writing–especially narrative fiction and writing for film–and environmental literature. Now he devotes most of his time to projects designed to enhance the farm and support the vision of Treleven, Inc.

Chelsea Green Publishing is committed to preserving ancient forests and natural resources. We elected to print this title on 30-percent postconsumer recycled paper, processed chlorine-free. As a result, for this printing, we have saved:

10 Trees (40' tall and 6-8" diameter)
4 Million BTUs of Total Energy
846 Pounds of Greenhouse Gases
4,589 Gallons of Wastewater
307 Pounds of Solid Waste

Chelsea Green Publishing made this paper choice because we and our printer, Thomson-Shore, Inc., are members of the Green Press Initiative, a nonprofit program dedicated to supporting authors, publishers, and suppliers in their efforts to reduce their use of fiber obtained from endangered forests. For more information, visit: www.greenpressinitiative.org.

Environmental impact estimates were made using the Environmental Defense Paper Calculator. For more information visit: www.papercalculator.org.